FUNDAMENTALS OF
DNA METHYLATION

FUNDAMENTALS OF DNA METHYLATION

Dr. R.N. Shukla

Oxford Book Company
Jaipur • New Delhi

Published By

Oxford Book Company

Regd. Off. "Bony Residency", Gate No. 2
Opp. Tilak Public School, Vishweswariya Nagar,
Gopalpura Road, Jaipur - 302018, India
Ph: 0141 - 2761280, Telefax: 2761381
email - oxfordbookjpr@gmail.com
website - oxfordbookcompany.com

Delhi Office: 102, 1st Floor, Satyam House,
4327/3, Ansari Road, Darya Ganj,
New Delhi - 110002, India
Ph: 011-45652440

Distributed By

KEYA INTERNATIONAL PRESS LLP
E-mail: keyainternationalpressllp@gmail.com

Published in India, 2025

ISBN : 9789391096502

Typeset by Shivangi Computers, Jaipur

Printed and bound in India by Thomson Press Ltd.,
New Delhi

Contents

Preface

DNA usually exists as a double-stranded structure, with both strands coiled together to form the characteristic double-helix. Each single strand of DNA is a chain of four types of nucleotides having the bases: adenine, cytosine, guanine, and thymine. A nucleotide is a mono-, di-, or triphosphate deoxyribonucleoside; that is, a deoxyribose sugar is attached to one, two, or three phosphates. Chemical interaction of these nucleotides forms phosphodiester linkages, creating the phosphate-deoxyribose backbone of the DNA double helix with the bases pointing inward. Nucleotides are matched between strands through hydrogen bonds to formbase pairs. Adenine pairs with thymine, and cytosine pairs with guanine.

DNA strands have a directionality, and the different ends of a single strand are called the "3' end" and the "5' end" with the direction of the naming going 5 prime to the 3 prime region.The strands of the helix are anti-parallel with one being 5 prime to 3 then the opposite strand 3 prime to 5 . These terms refer to the carbon atom in deoxyribose to which the next phosphate in the chain attaches. In addition to being complementary, the two strands of DNA are antiparallel: They are orientated in opposite directions. This directionality has consequences in DNA synthesis, because DNA polymerase can synthesize DNA in only one direction by adding nucleotides to the 3' end of a DNA strand. The pairing of bases in DNA through hydrogen bonding means that the information contained within each strand is redundant. The nucleotides on a single strand can be used to reconstruct nucleotides on a newly synthesized partner strand.

DNA polymerases adds nucleotides to the 5' end of a strand of DNA. If a mismatch is accidentally incorporated, the polymerase is inhibited from further extension. Proofreading removes the mismatched nucleotide and extension continues. DNA polymerases are a family of enzymes that carry out all forms of DNA replication. However, a DNA polymerase can only extend an existing DNA strand paired with a template strand; it cannot begin the synthesis of a new strand. To begin synthesis, a short fragment of DNA or RNA, called a primer, must be created and paired with the template DNA strand.

DNA polymerase then synthesizes a new strand of DNA by extending the 3' end of an existing nucleotide chain, adding new nucleotides matched to the template strand one at a time via the creation of phosphodiester bonds. The energy for this process of DNA polymerization comes from two of the three total phosphates attached to each unincorporated base. When a nucleotide is being added to a growing DNA strand, two of the phosphates are removed and the energy produced creates a phosphodiester bond that attaches the remaining phosphate to the growing chain. The energetics of this process also help explain the directionality of synthesis - if DNA were synthesized in the 3' to 5' direction, the energy for the process would come from the 5' end of the growing strand rather than from free nucleotides.

The book will convey the excitement of the continuing process of discoveries and the strategies developing in genetic engineering.

— *Dr. R.N. Shukla*

Chapter 1

Methylation of RNA

Variation in methylation states of DNA can alter gene expression levels significantly. Methylation variation usually occurs through the action of DNA methylases. When the change is heritable, it is considered epigenetic. When the change in information status is not heritable, it would be a somatic epitype. The effective information content has been changed by means of the actions of a protein or proteins on DNA, but the primary DNA sequence is not altered.

Prions

Prions are proteins that propagate themselves by making conformational changes in other molecules of the same type of protein. This change affects the behaviour of the protein. In fungi this change happens from one generation to the next, i.e. Protein '! Protein. While this represents a transfer of information, prion interactions leave the sequence of the protein unchanged, and so are not technically considered an exception to the central dogma.

Use of the Term "Dogma"

In his autobiography, What Mad Pursuit, Crick wrote about his choice of the word dogma and some of the problems it caused him:

"I called this idea the central dogma, for two reasons, I suspect. I had already used the obvious word hypothesis in the sequence hypothesis, and in addition I wanted to suggest that this new assumption was more central and more powerful. ... As it turned out, the use of the word dogma caused almost more trouble than it was worth.... Many years later Jacques Monodpointed out to me that I did not appear to understand the correct use of the word dogma, which is a belief that cannot be doubted. I did apprehend this in a vague sort of way but since I thought that all religious beliefs were without foundation, I used the word the way I myself thought about it, not as most of the world does, and simply applied it to a grand hypothesis that, however plausible, had little direct experimental support."

"My mind was, that a dogma was an idea for which there was no reasonable evidence. You see?!" And Crick gave a roar of delight. "I just didn't know what dogma meant. And I could just as well have called it the 'Central Hypothesis,' or — you know. Which is what I meant to say. Dogma was just a catch phrase."

Translation of mRNA

Proteins are synthesized from mRNA templates by a process that has been highly conserved throughout evolution (reviewed in Chapter 3). All mRNAs are read in the 5′ to 3′ direction, and polypeptide chains are synthesized from the amino to the carboxy terminus. Each amino acid is specified by three bases (a codon) in the mRNA, according to a nearly universal genetic code. The basic mechanics of protein synthesis are also the same in all cells: Translation is carried out on ribosomes, with tRNAs serving as adaptors between the mRNA template and the amino acids being incorporated into protein. Protein synthesis thus involves interactions between three types of RNA molecules (mRNA templates, tRNAs, and rRNAs), as well as various proteins that are required for translation.

Transfer RNAs

During translation, each of the 20 amino acids must be aligned with their corresponding codons on the mRNA template. All cells contain a variety of tRNAs that serve as adaptors for this process. As might be expected, given their common function in protein synthesis, different tRNAs share similar overall structures. However, they also possess unique identifying sequences that allow the correct amino acid to be attached and aligned with the appropriate codon in mRNA.

Transfer RNAs are approximately 70 to 80 nucleotides long and have characteristic cloverleaf structures that result from complementary base pairing between different regions of the molecule. X-ray crystallography studies have further shown that all tRNAs fold into similar compact L shapes, which are likely required for the tRNAs to fit onto ribosomes during the translationprocess. The adaptor function of the tRNAs involves two separated regions of the molecule. All tRNAs have the sequence CCA at their 3′ terminus, and amino acids are covalently attached to the ribose of the terminal adenosine. The mRNA template is then recognized by the anticodon loop, located at the other end of the folded tRNA, which binds to the appropriate codon by complementary base pairing.

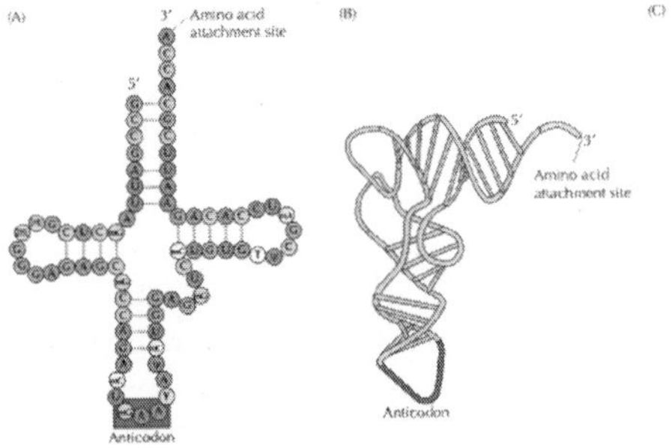

Fig. 1:

Structure of tRNAs. The structure of yeast phenylalanyl tRNA is illustrated in open "cloverleaf" form (A) to show complementary base pairing. Modified bases are indicated as mG, methylguanosine; mC, methylcytosine; DHU, dihydrouridine.

The incorporation of the correctly encoded amino acids into proteins depends on the attachment of each amino acid to an appropriate tRNA, as well as on the specificity of codon-anticodon base pairing. The attachment of amino acids to specific tRNAs is mediated by a group of enzymes called aminoacyl tRNA synthetases, which were discovered by Paul Zamecnik and Mahlon Hoagland in 1957.

Each of these enzymes recognizes a single amino acid, as well as the correct tRNA (or tRNAs) to which that amino acid should be attached. The reaction proceeds in two steps. First, the amino acid is activated by reaction with ATP to form an aminoacyl AMP synthetase intermediate.

The activated amino acid is then joined to the 3′ terminus of the tRNA. The aminoacyl tRNA synthetases must be highly selective enzymes that recognize both individual amino acids and specific base sequences that identify the correct acceptor tRNAs. In some cases, the high fidelity of amino acid recognition results in part from a proofreadingfunction by which incorrect aminoacyl AMPs are hydrolyzed rather than being joined to tRNA during the second step of the reaction.

Recognition of the correct tRNA by the aminoacyl tRNA synthetase is also highly selective; the synthetase recognizes specificnucleotide sequences (in most cases including the anticodon) that uniquely identify each species of tRNA.

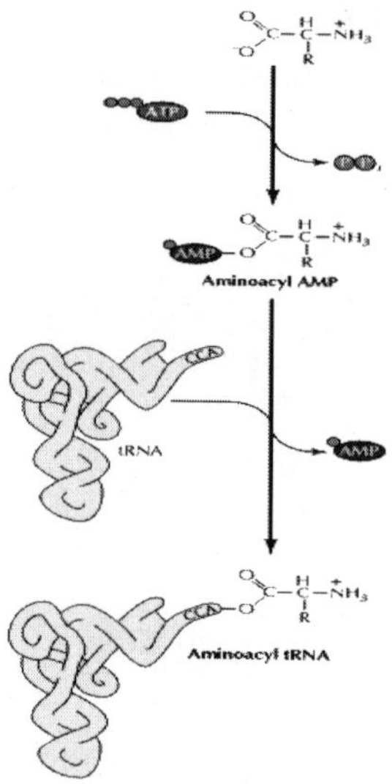

Fig. 2: Attachment of amino acids to tRNAs. In the first reaction step, the amino acid is joined to AMP, forming an aminoacyl AMP intermediate. In the second step, the amino acid is transferred to the 3′ CCA terminus of the acceptor tRNA and AMP is released.

After being attached to tRNA, an amino acid is aligned on the mRNA template by complementary base pairing between the mRNAcodon and the anticodon of the tRNA. Codon-anticodon base pairing is somewhat less stringent than the standard A-U and G-C base pairing discussed in preceding chapters. The significance of this unusual base pairing in codon-anticodon recognition relates to the redundancy of the genetic code. Of the 64 possible codons, three are stop codons that signal the termination of translation; the other 61 encode amino acids.

Thus, most of the amino acids are specified by more than one codon. In part, this redundancy results from the attachment of many amino acids to more than one species of tRNA. E. coli, for example, contain about 40 different tRNAs that serve as acceptors for the 20 different amino acids.

In addition, some tRNAs are able to recognize more than one codon in mRNA, as a result of nonstandard base pairing (called wobble) between the tRNA anticodon and the third position of some complementary codons. Relaxed base pairing at this position results partly from the formation of G-U base pairs and partly from the modification of guanosine to inosine in the anticodons of several tRNAs during processing. Inosine can base-pair with either C, U, or A in the third position, so its inclusion in the anticodon allows a single tRNA to recognize three different codons in mRNA templates.

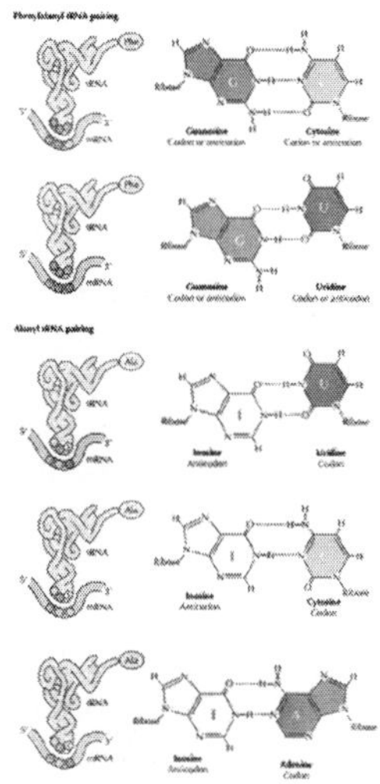

Fig. 3: Nonstandard codon-anticodon base pairing. Base pairing at the third codon position is relaxed, allowing G to pair with U, and inosine (I) in the anticodon to pair with U, C, or A. Two examples of abnormal base pairing, allowing phenylalanyl (Phe) tRNA.

The Ribosome

Ribosomes are the sites of protein synthesis in both prokaryotic and eukaryotic cells. First characterized as particles detected by ultracentrifugation of cell lysates, ribosomes are usually designated according to their rates of sedimentation: 70S for bacterialribosomes and 80S for the somewhat larger ribosomes of eukaryotic cells. Both prokaryotic and eukaryotic ribosomes are composed of two distinct subunits, each containing characteristic proteins and rRNAs. The fact that cells typically contain many ribosomesreflects the central importance of protein synthesis in cell metabolism. E. coli, for example, contain about 20,000 ribosomes, which account for approximately 25% of the dry weight of the cell, and rapidly growing mammalian cells contain about 10 million ribosomes.

The general structures of prokaryotic and eukaryotic ribosomes are similar, although they differ in some details. The small subunit (designated 30S) of E. coli ribosomes consists of the 16S rRNA and 21 proteins; the large subunit (50S) is composed of the 23S and 5S rRNAs and 34 proteins. Each ribosome contains one copy of the rRNAs and one copy of each of the ribosomalproteins, with one exception: One protein of the 50S subunit is present in four copies. The subunits of eukaryotic ribosomes are larger and contain more proteins than their prokaryotic counterparts have. The small subunit (40S) of eukaryotic ribosomes is composed of the 18S rRNA and approximately 30 proteins; the large subunit (60S) contains the 28S, 5.8S, and 5S rRNAs and about 45 proteins.

Ribosome structure. (A) Electron micrograph of E. coli 50S ribosomal subunits. (B–C) High resolution X-ray crystal structures of 30S (B) and 50S (C) ribosomal subunits. (D) Model of ribosome structure. (E) Components of prokaryotic and eukaryotic.

A noteworthy feature of ribosomes is that they can be formed in vitro by self-assembly of their RNA and protein constituents. As first described in 1968 by Masayasu Nomura, purified ribosomal proteins and rRNAs can be mixed together and, under appropriate conditions, will reform a functional ribosome. Although ribosome assembly in vivo (particularly in eukaryotic cells) is considerably more complicated, the ability of ribosomes to self-assemble in vitro has provided an important experimental tool, allowing analysis of the roles of individual proteins and rRNAs.

Like tRNAs, rRNAs form characteristic secondary structures by complementary base pairing. In association with ribosomal proteins the rRNAs fold further, into distinct three-dimensional structures. Initially, rRNAs were thought to play a structural role, providing a scaffold upon which ribosomal proteins assemble. However, with the discovery of the catalytic activity of other RNA molecules (e.g., RNase P and the self-splicing introns discussed in Chapter 6), the possible catalytic role of rRNA became widely considered. Consistent with this hypothesis, rRNAs were found to be absolutely required for the in vitro assembly of functionalribosomes. On the other hand, the omission of many ribosomal proteins resulted in a decrease, but not a complete loss, of ribosome activity.

The structure of 16S rRNA. Complementary base pairing results in the formation of a distinct secondary structure. (From S. Stern, T. Powers, L.-I. Changchien and H. F. Noller, 1989. Science 244: 783.)

Direct evidence for the catalytic activity of rRNA first came from experiments of Harry Noller and his colleagues in 1992. These investigators demonstrated that the large ribosomal subunit is able to catalyze the formation of peptide bonds (the peptidyl transferase reaction) even after approximately 95% of the ribosomal proteins have been removed by standard protein extraction procedures. In contrast, treatment with RNase completely abolishes peptide bond formation, providing strong support for the hypothesis that the formation of a peptide bond is an RNA-catalyzed reaction. Further studies have confirmed and extended these results by demonstrating that the peptidyl transferase reaction can be catalyzed by synthetic fragments of 23S rRNA in the total absence of any ribosomal protein. Thus, the fundamental reaction of protein synthesis is catalyzed by ribosomal RNA. Rather than being the primary catalytic constituents of ribosomes, ribosomal proteins are now thought to facilitate proper folding of the rRNA and to enhance ribosome function by properly positioning the tRNAs.

The direct involvement of rRNA in the peptidyl transferase reaction has important evolutionary implications. RNAs are thought to have been the first self-replicating macromolecules. This notion is strongly supported by the fact that ribozymes, such asRNase P and self-splicing introns, can catalyze reactions that involve RNA substrates. The role of rRNA in the formation of peptide bonds extends the catalytic activities of RNA beyond self-replication to direct involvement in protein synthesis. Additional studies indicate that the Tetrahymena rRNA ribozyme can catalyze the attachment of amino acids to RNA, lending credence to the possibility that the original aminoacyl tRNA synthetases were RNAs rather than proteins. The ability of RNA molecules to catalyze the reactions required for protein synthesis as well as for self-replication may provide an important link for understanding the early evolution of cells.

The Organization of mRNAs and the Initiation of Translation

Although the mechanisms of protein synthesis in prokaryotic and eukaryotic cells are similar, there are also differences, particularly in the signals that determine the positions at which synthesis of a polypeptide chain is initiated on an mRNA template.Translation does not simply begin at the 5′ end of the mRNA; it starts at specific initiation sites. The 5′ terminal portions of both prokaryotic and eukaryotic mRNAs are therefore noncoding sequences, referred to as 5′ untranslated regions. Eukaryotic mRNAs usually encode only a single polypeptide chain, but many prokaryotic mRNAs encode multiple polypeptides that are synthesized independently from distinct initiation sites. For example, the E. coli lac operon consists of three genes that are translated from the same mRNA. Messenger RNAs that encode multiple polypeptides are called polycistronic, whereas monocistronicmRNAs encode a single polypeptide chain. Finally, both prokaryotic and eukaryotic mRNAs end in noncoding 3′ untranslated regions.

Prokaryotic and eukaryotic mRNAs. Both prokaryotic and eukaryotic mRNAs contain untranslated regions (UTRs) at their 5′ and 3′ ends. Eukaryotic mRNAs also contain 5′ 7-methylguanosine (mG) caps and 3′ poly-A tails. Prokaryotic.

In both prokaryotic and eukaryotic cells, translation always initiates with the amino acid methionine, usually encoded by AUG. Alternative initiation codons, such as GUG, are used occasionally in bacteria, but when they occur at the beginning of a polypeptidechain, these codons direct the incorporation of methionine rather than of the amino acid they normally encode (GUG normally encodes valine). In most bacteria, protein synthesis is initiated with a modified methionine residue (N-formylmethionine), whereas unmodified methionines initiate protein synthesis in eukaryotes (except in mitochondria and chloroplasts, whose ribosomes resemble those of bacteria).

The signals that identify initiation codons are different in prokaryotic and eukaryotic cells, consistent with the distinct functions ofpolycistronic and monocistronic mRNAs. Initiation codons in bacterial mRNAs are preceded by a specific sequence (called a Shine-Delgarno sequence, after its discoverers) that aligns the mRNA on the ribosome for translation by base-pairing with a complementary sequence near the 3′ terminus of 16S rRNA. This base-pairing interaction enables bacterial ribosomes to initiatetranslation not only at the 5′ end of an mRNA but also at the internal initiation sites of polycistronic messages. In contrast,ribosomes recognize most eukaryotic mRNAs by binding to the 7-methylguanosine cap at their 5′ terminus. Theribosomes then scan downstream of the 5′ cap until they encounter an AUG initiation codon. Sequences that surround AUGs affect the efficiency of initiation, so in many cases the first AUG in the mRNA is bypassed and translation initiates at an AUG farther downstream. However, eukaryotic mRNAs have no sequence equivalent to the Shine-Delgarno sequence of prokaryotic mRNAs.Translation of eukaryotic mRNAs is instead initiated at a site determined by scanning from the 5′ terminus, consistent with their functions as monocistronic messages that encode only single polypeptides.

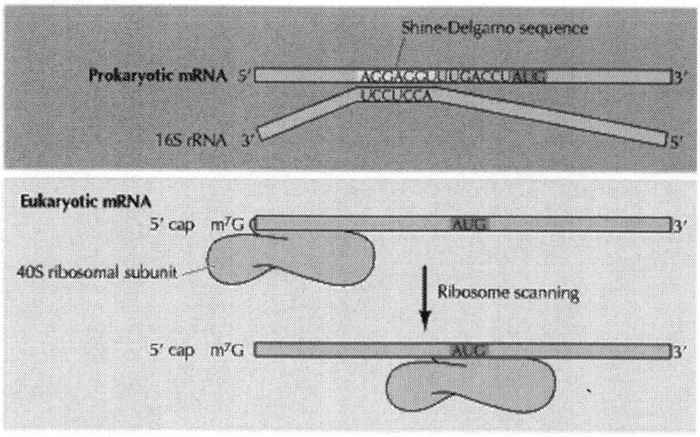

Fig. 4:

Signals for translation initiation. Initiation sites in prokaryotic mRNAs are characterized by a Shine-Delgarno sequence that precedes the AUG initiation codon. Base pairing between the Shine-Delgarno sequence and a complementary sequence near the 3′.

The Process of Translation

Translation is generally divided into three stages: initiation, elongation, and termination. In both prokaryotes and eukaryotes the first step of the initiation stage is the binding of a specific initiator methionyl tRNA and the mRNA to the small ribosomal subunit. The large ribosomal subunit then joins the complex, forming a functional ribosome on which elongation of thepolypeptide chain proceeds. A number of specific nonribosomal proteins are also required for the various stages of the translationprocess.

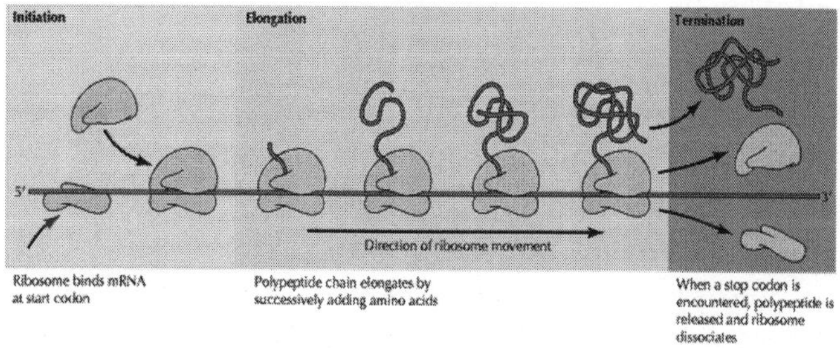

Fig. 5:

Translation Factors

The first translation step in bacteria is the binding of three initiation factors (IF-1, IF-2, and IF-3) to the 30S ribosomal subunit.

The mRNA and initiator N-formylmethionyl tRNA then join the complex, with IF-2 (which is bound to GTP) specifically recognizing the initiator tRNA. IF-3 is then released, allowing a 50S ribosomal subunit to associate with the complex. This association triggers the hydrolysis of GTP bound to IF-2, which leads to the release of IF-1 and IF-2 (bound to GDP). The result is the formation of a 70S initiation complex (with mRNA and initiator tRNA bound to the ribosome) that is ready to begin peptide bondformation during the elongation stage of translation.

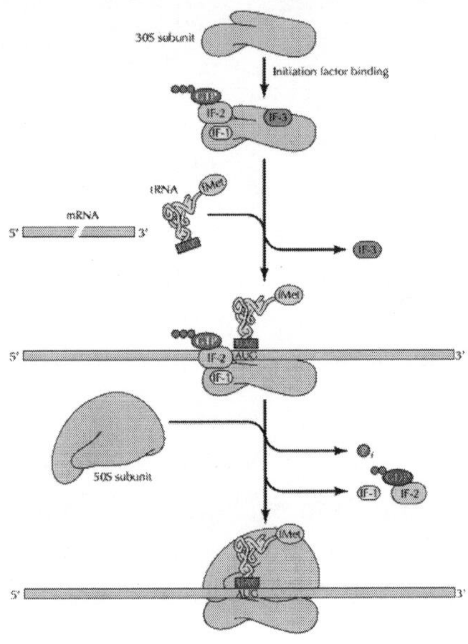

Fig. 6.

Initiation of translation in bacteria. Three initiation factors (IF-1, IF-2, and IF-3) first bind to the 30S ribosomal subunit. This step is followed by binding of the mRNA and the initiatorN-formylmethionyl (fMet) tRNA, which is recognized by IF-2 bound.

Initiation in eukaryotes is more complicated and requires at least ten proteins (each consisting of multiple polypeptide chains), which are designated eIFs. The factors eIF-1, eIF-1A, and eIF-3 bind to the 40S ribosomal subunit, and eIF-2 (in a complex with GTP) associates with the initiator methionyl tRNA. The mRNA is recognized and brought to the ribosome by the eIF-4 group of factors. The 5′ cap of the mRNA is recognized by eIF-4E. Another factor, eIF-4G, binds to both eIF-4E and to a protein (poly-A binding protein or PABP) associated with the poly-A tail at the 3' end of the mRNA. Eukaryotic initiation factors thus recognize both the 5' and 3' ends of mRNAs, accounting for the stimulatory effect of polyadenylationon translation. The initiation factors eIF-4E and eIF-4G, in association with eIF-4A and eIF-4B, then bring the mRNA to the 40S ribosomal subunit, with eIF-4G interacting with eIF-3. The 40S ribosomal subunit, in association with the bound methionyl tRNA and eIFs, then scans the mRNA to identify the AUG initiation codon. When the AUG codon is reached, eIF-5 triggers the hydrolysis of GTP bound to eIF-2. Initiation factors (including eIF-2 bound to GDP) are then released, and a 60S subunit binds to the 40S subunit to form the 80S initiation complex of eukaryotic cells.

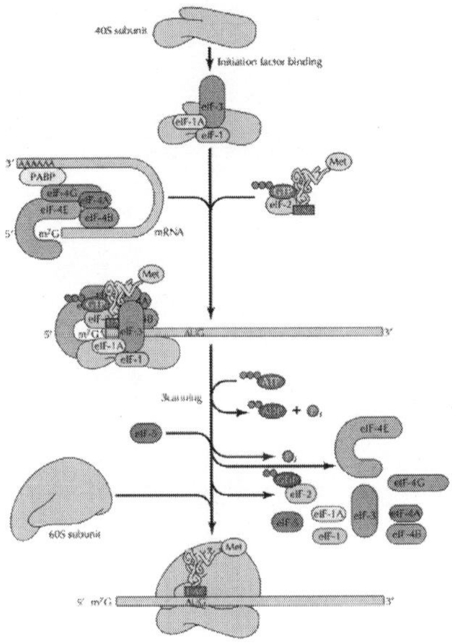

Fig. 7:

Initiation of translation in eukaryotic cells. Initiation factors eIF-3, eIF-1, and eIF-1A bind to the 40S ribosomal subunit. The initiator methionyl tRNA is brought to the ribosome by eIF-2 (complexed to GTP), and the mRNA by eIF-4E.

After the initiation complex has formed, translation proceeds by elongation of the polypeptide chain. The mechanism of elongation in prokaryotic and eukaryotic cells is very similar. The ribosome has three sites for tRNA binding, designated the P (peptidyl), A (aminoacyl), and E (exit) sites. The initiator methionyl tRNA is bound at the P site. The first step in elongation is the binding of the next aminoacyl tRNA to the A site by pairing with the second codon of the mRNA. The aminoacyl tRNA is escorted to the ribosome by an elongation factor

(EF-Tu in prokaryotes, eEF-1± in eukaryotes), which is complexed to GTP. The GTP is hydrolyzed to GDP as the correct aminoacyl tRNA is inserted into the A site of the ribosome and the elongation factor bound to GDP is released. The requirement for hydrolysis of GTP before EF-Tu or eEF-1± is released from the ribosome is the rate-limiting step in elongation and provides a time interval during which an incorrect aminoacyl tRNA, which would bind less strongly to the mRNAcodon, can dissociate from the ribosome rather than being used for protein synthesis. Thus, the expenditure of a high-energy GTP at this step is an important contribution to accurate protein synthesis; it allows time for proofreading of the codon-anticodon pairing before the peptide bond forms.

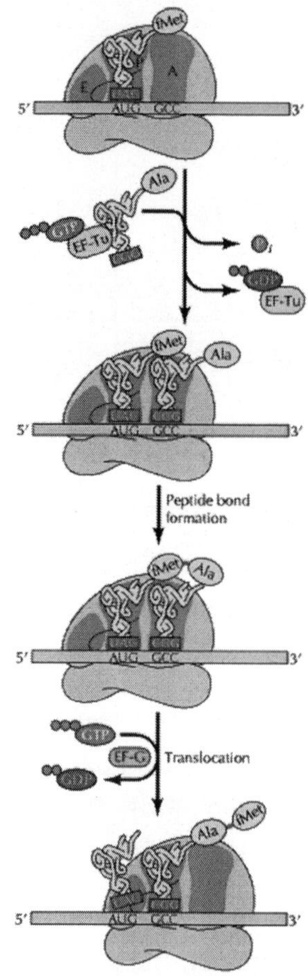

Fig. 8:

Elongation stage of translation. The ribosome has three tRNA-binding sites, designated P (peptidyl), A (aminoacyl), and E (exit). The initiating N-formylmethionyl tRNA is positioned in the P site, leaving an empty A site. The second aminoacyl tRNA.

Once EF-Tu (or eEF-1±) has left the ribosome, a peptide bond can be formed between the initiator methionyl tRNA at the P site and the second aminoacyl tRNA at the A site. This reaction is catalyzed by the large ribosomal subunit, with the rRNA playing a critical role.

The result is the transfer of methionine to the aminoacyl tRNA at the A site of the ribosome, forming a peptidyl tRNA at this position and leaving the uncharged initiator tRNA at the P site. The next step in elongation is translocation, which requires another elongation factor (EF-G in prokaryotes, eEF-2 in eukaryotes) and is again coupled to GTP hydrolysis.

During translocation, the ribosome moves three nucleotides along the mRNA, positioning the next codon in an empty A site. This step translocates the peptidyl tRNA from the A site to the P site, and the uncharged tRNA from the P site to the E site.

The ribosome is then left with a peptidyl tRNA bound at the P site, and an empty A site. The binding of a new aminoacyl tRNA to the A site then induces the release of the uncharged tRNA from the E site, leaving the ribosome ready for insertion of the next amino acid in the growing polypeptide chain.

As elongation continues, the EF-Tu (or eEF-1±) that is released from the ribosome bound to GDP must be reconverted to its GTP form. This conversion requires a third elongation factor, EF-Ts (eEF-1[23] in eukaryotes), which binds to the EF-Tu/GDP complex and promotes the exchange of bound GDP for GTP.

This exchange results in the regeneration of EF-Tu/GTP, which is now ready to escort a new aminoacyl tRNA to the A site of the ribosome, beginning a new cycle of elongation. The regulation of EF-Tu by GTP binding and hydrolysis illustrates a common means of the regulation of protein activities.

As will be discussed in later chapters, similar mechanisms control the activities of a wide variety of proteins involved in the regulation of cell growth and differentiation, as well as in protein transport and secretion.

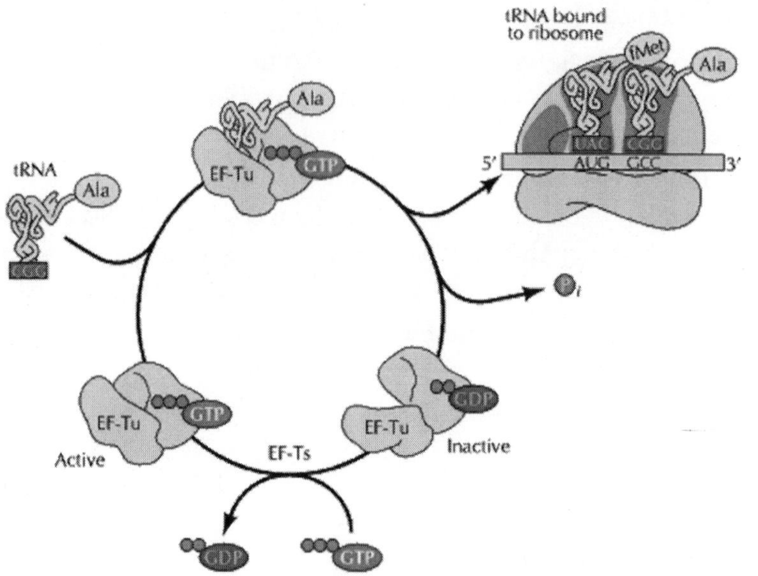

Fig. 9:

Regeneration of EF-Tu/GTP. EF-Tu complexed to GTP escorts the aminoacyl tRNA to the ribosome. The bound GTP is hydrolyzed as the correct tRNA is inserted, so EF-Tu complexed to GDP is released. The EF-Tu/GDP complex is inactive and unable to bind another.

Elongation of the polypeptide chain continues until a stop codon (UAA, UAG, or UGA) is translocated into the A site of the ribosome. Cells do not contain tRNAs with anticodons complementary to these termination signals; instead, they have release factors that recognize the signals and terminate protein synthesis.

Prokaryotic cells contain two release factors that recognize termination codons: RF-1 recognizes UAA or UAG, and RF-2 recognizes UAA or UGA. In eukaryotic cells a single release factor (eRF-1) recognizes all three termination codons.

Both prokaryotic and eukaryotic cells also contain release factors (RF-3 and eRF-3, respectively) that do not recognize specific termination codons but act together with RF-1 (or eRF-1) and RF-2. The release factors bind to a termination codon at the A site and stimulate hydrolysis of the bond between the tRNA and the polypeptidechain at the P site, resulting in release of the completed polypeptide from the ribosome. The tRNA is then released, and the ribosomal subunits and the mRNA template dissociate.

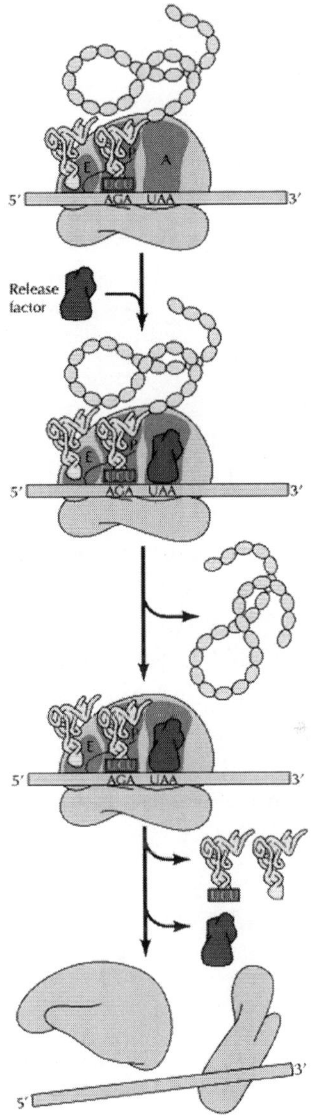

Fig. 10.

Termination of translation. A termination codon (e.g., UAA) at the A site is recognized by a release factor rather than by a tRNA. The result is the release of the completed polypeptide chain, followed by the dissociation of tRNA and mRNA from the ribosome.

Messenger RNAs can be translated simultaneously by several ribosomes in both prokaryotic and eukaryotic cells. Once one ribosome has moved away from the initiation site, another can bind to the mRNA and begin synthesis of a new polypeptide chain. Thus, mRNAs are usually translated by a series of ribosomes, spaced at intervals of about 100 to 200 nucleotides. The group of ribosomes bound to an mRNA molecule is called a polyribosome, or polysome. Each ribosome within the group functions independently to synthesize a separate polypeptide chain.

Polysomes. Messenger RNAs are translated by a series of multiple ribosomes (a polysome). (A) Electron micrograph of a eukaryotic polysome. (B) Schematic of a generalized poly-some.

Regulation of Translation

Although transcription is the primary level at which gene expression is controlled, the translation of mRNAs is also regulated in both prokaryotic and eukaryotic cells. One mechanism of translational regulation is the binding of repressor proteins, which blocktranslation, to specific mRNA sequences. The best understood example of this mechanism in eukaryotic cells is regulation of the synthesis of ferritin, a protein that stores iron within the cell. The translation of ferritin mRNA is regulated by the supply of iron: More ferritin is synthesized if iron is abundant. This regulation is mediated by a protein which (in the absence of iron) binds to a sequence (the iron response element, or IRE) in the 5′ untranslated region of ferritin mRNA, blocking its translation. In the presence of iron, the repressor no longer binds to the IRE and ferritin translation is able to proceed.

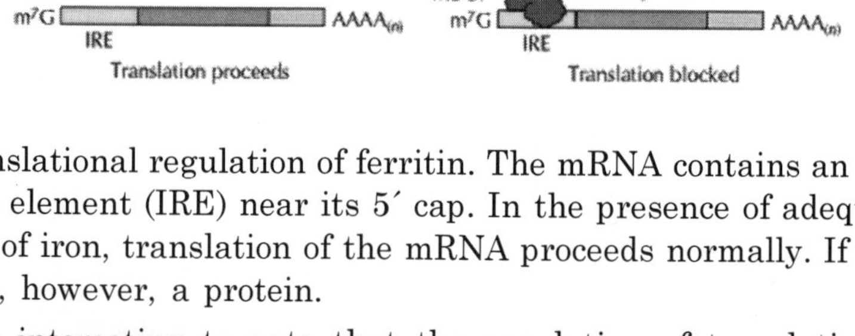

Fig. 11:

Translational regulation of ferritin. The mRNA contains an iron response element (IRE) near its 5′ cap. In the presence of adequate supplies of iron, translation of the mRNA proceeds normally. If iron is scarce, however, a protein.

It is interesting to note that the regulation of translation of ferritin mRNA by iron is similar to the regulation of transferrin receptor mRNA stability, which was discussed in the previous chapter. Namely, the stability of transferrin receptor mRNA is regulated by protein binding to an IRE in its 3′ untranslated region. The same protein binds to the IREs of both ferritin and transferrin receptor mRNAs. However, the consequences of protein binding to the two IREs are quite different. Protein bound to the transferrin receptor IRE protects the mRNA from degradation rather than inhibiting its translation.

These distinct effects presumably result from the different locations of the IRE in the two mRNAs. To function as a repressor-binding site, the IRE must be located within 70nucleotides of the 5′ cap of ferritin mRNA, suggesting that protein binding to the IRE blocks translation by interfering with cap recognition and binding of the 40S ribosomal subunit. Rather than inhibiting translation, protein binding to the same sequence in the 3′ untranslated region of transferrin receptor mRNA protects the mRNA from nuclease degradation. Binding of the same regulatory protein to different sites on mRNA molecules can thus have distinct effects on gene expression, in one case inhibiting translation and in the other stabilizing the mRNA to increase protein synthesis.

Another mechanism of translational regulation in eukaryotic cells, resulting in global effects on overall translational activity rather than on the translation of specific mRNAs, involves modulation of the activity of initiation factors, particularly eIF-2. As already discussed, eIF-2 (complexed with GTP) binds to the initiator methionyl tRNA, bringing it to the ribosome. The subsequent release of eIF-2 is accompanied by GTP hydrolysis, leaving eIF-2 as an inactive GDP complex. To participate in another cycle of initiation, the eIF-2/GTP complex must be regenerated by the exchange of bound GDP for GTP. This exchange is mediated by another factor, eIF-2B.

The control of eIF-2 activity by GTP binding and hydrolysis is thus similar to that of EF-Tu. However, the regulation of eIF-2 provides a critical control point in a variety of eukaryotic cells. In particular, eIF-2 can be phosphorylated by regulatory protein kinases. This phosphorylation blocks the exchange of bound GDP for GTP, thereby inhibiting initiation of translation. One type of cell in which such phosphorylation occurs is the reticulocyte, which is devoted to the synthesis of hemoglobin. Thetranslation of globin mRNA is controlled by the availability of heme: The mRNA is translated only if adequate heme is available to form functional hemoglobin molecules. In the absence of heme, a protein kinase that phosphorylates eIF-2 is activated, and furthertranslation is inhibited. Similar mechanisms have been found to control protein synthesis in other cell types, particularly virus-infected cells in which viral protein synthesis is inhibited by interferon.

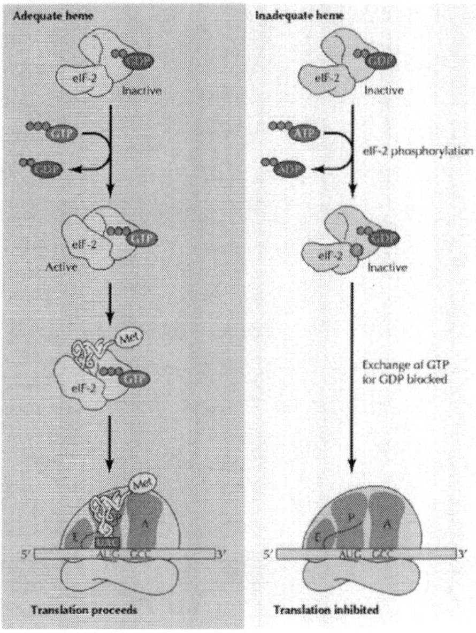

Fig. 12:

Regulation of translation by phosphorylation of eIF-2. Translation in reticulocytes (which is devoted to synthesis of hemoglobin) is controlled by the supply of heme, which regulates the activity of eIF-2. The active form of eIF-2.

Other studies have implicated eIF-4E, which binds to the 5′ cap of mRNAs, as a translational regulatory protein. For example, the hormone insulin stimulates protein synthesis in adipocytes and muscle cells. This effect of insulin is mediated, at least in part, byphosphorylation of proteins associated with eIF-4E, resulting in stimulation of eIF-4E activity and increased rates of translational initiation.

Translational regulation is particularly important during early development. As discussed in Chapter 6, a variety of mRNAs are stored in oocytes in an untranslated form; the translation of these stored mRNAs is activated at fertilization or later stages of development. One mechanism of such translational regulation is the controlled polyadenylation of oocyte mRNAs. Many untranslated mRNAs are stored in oocytes with short poly-A tails (approximately 20 nucleotides). These stored mRNAs are subsequently recruited fortranslation at the appropriate stage of development by the lengthening of their poly-A tails to several hundred nucleotides. In addition, the translation of some mRNAs during development appears to be regulated by repressor proteins that bind to specific sequences in their 3′ untranslated regions. These regulatory proteins may also direct mRNAs to specific regions of eggs or embryos, allowing localized synthesis of the encoded proteins during embryonic development.

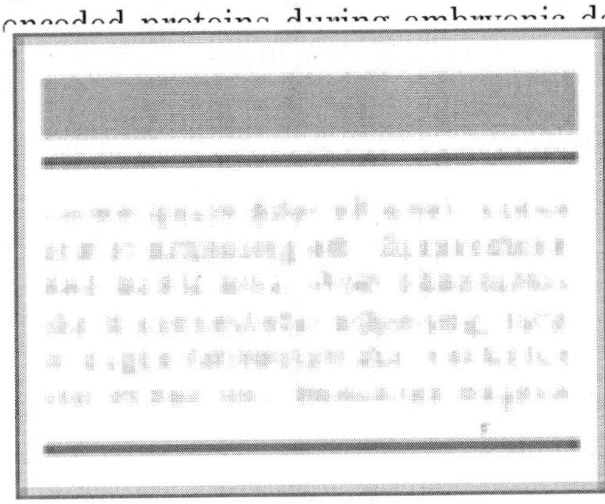

Fig. 13: Box

Key Experiment: Catalytic Role of Ribosomal RNA. Unusual Resistance of Peptidyl Transferase to Protein Extraction Procedures Harre F. Noller, Vernita Hoffarth, and Ludwika Zimniak

Fig. 14: Box

Molecular Medicine: Antibiotics and Protein Synthesis. Bacteria are responsible for a wide variety of potentially lethal infectious diseases, including tuberculosis, bacterial pneumonia, childhood meningitis, infections of wounds and burns.

By agreement with the publisher, this book is accessible by the search feature, but cannot be browsed.

Gene Expression: Transcription

The majority of genes are expressed as the proteins they encode. The process occurs in two steps:

- Transcription = DNA '! RNA
- Translation = RNA '! protein

Taken together, they make up the "central dogma" of biology: DNA → RNA → protein.

Here is an overview.

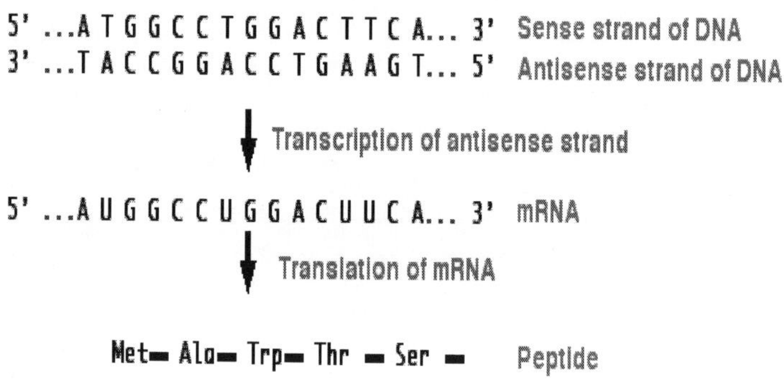

This page examines the first step:

Gene Transcription: DNA → RNA

DNA serves as the template for the synthesis of RNA much as it does for its own replication.

The Steps

- Some 50 different protein transcription factors bind to promoter sites, usually on the 52 side of the gene to be transcribed.
- An enzyme, an RNA polymerase, binds to the complex of transcription factors.
- Working together, they open the DNA double helix.
- The RNA polymerase proceeds to "read" one strand moving in its 32 '! 52 direction.

- In eukaryotes, this requires — at least for protein-encoding genes — that the nucleosomes in front of the advancing RNA polymerase (RNAP II) be removed. A complex of proteins is responsible for this. The same complex replaces the nucleosomes after the DNA has been transcribed and RNAP II has moved on.
- As the RNA polymerase travels along the DNA strand, it assembles ribonucleotides (supplied as triphosphates, e.g., ATP) into a strand of RNA.
- Each ribonucleotide is inserted into the growing RNA strand following the rules of base pairing. Thus for each C encountered on the DNA strand, a G is inserted in the RNA; for each G, a C; and for each T, an A. However, each A on the DNA guides the insertion of the pyrimidine uracil (U, from uridine triphosphate, UTP). There is no T in RNA.
- TABLE
- Synthesis of the RNA proceeds in the 52 '! 32 direction.
- As each nucleoside triphosphate is brought in to add to the 32 end of the growing strand, the two terminal phosphates are removed.
- When transcription is complete, the transcript is released from the polymerase and, shortly thereafter, the polymerase is released from the DNA.

Note that at any place in a DNA molecule, either strand may be serving as the template; that is, some genes "run" one way, some the other (and in a few remarkable cases, the same segment of double helix contains genetic information on both strands!).

In all cases, however, RNA polymerase transcribes the DNA strand in its 32 '! 52 direction.

A report in the 4 January 2001 issue of Nature shows that RNA polymerase actually tracks around the double helix of DNA. In vitro, at least, when RNA polymerase is immobilized, it spins the DNA molecule around and around as it moves along the molecule.

Whether it is the polymerase or the DNA that does the spinning in vivo remains to be determined.

Types of RNA

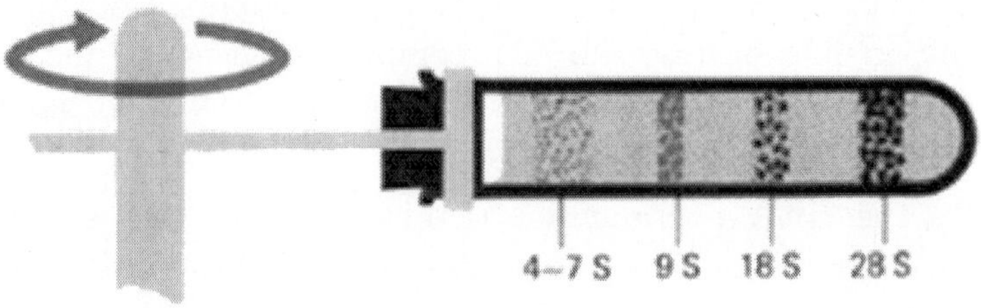

Fig. 15:

Sedimentation pattern produced by high-speed centrifugation of RNA extracted from the precursors of rabbit red blood cells. The discrete bands represent particular classes of RNA.

The transfer RNAs band at about 4S. The ribosomal RNAs of eukaryotes sediment at 5S, 5.8S, 18S, and 28S. (The larger the sedimentation unit, S, the larger the molecule — but not proportionally.)

The RNA forming the band at 9S is the messenger RNA for the synthesis of hemoglobin, the major protein synthesized in these cells. In most types of cells, the messenger RNAs are extremely heterogenous, with small amounts distributed from 6S to 25S.

Several types of RNA are synthesized in the nucleus of eukaryotic cells. Of particular interest are:

- messenger RNA (mRNA). This will later be translated into a polypeptide.
- ribosomal RNA (rRNA). This will be used in the building of ribosomes: machinery for synthesizing proteins by translating mRNA.
- transfer RNA (tRNA). RNA molecules that carry amino acids to the growing polypeptide.
- small nuclear RNA (snRNA). DNA transcription of the genes for mRNA, rRNA, and tRNA produces large precursor molecules ("primary transcripts") that must be processed within the nucleus to produce the functional molecules for export to the cytosol. Some of these processing steps are mediated by snRNAs.
- small nucleolar RNA (snoRNA). These RNAs within the nucleolus have several functions (described below).

- microRNA (miRNA). These are tiny (~22 nucleotides) RNA molecules that regulate the expression of messenger RNA (mRNA) molecules.
- XIST RNA. This inactivates one of the two X chromosomes in female vertebrates.

Messenger RNA (mRNA)

Messenger RNA comes in a wide range of sizes reflecting the size of the polypeptide it encodes. Most cells produce small amounts of thousands of different mRNA molecules, each to be translated into a peptide needed by the cell.

Many mRNAs are common to most cells, encoding "housekeeping" proteins needed by all cells (e.g., the enzymes of glycolysis). Other mRNAs are specific for only certain types of cells. These encode proteins needed for the function of that particular cell (e.g., the mRNA for hemoglobin in the precursors of red blood cells).

Ribosomal RNA (rRNA)

There are 4 kinds. In eukaryotes, these are

- 18S rRNA. One of these molecules, along with some 30 different protein molecules, is used to make the small subunit of the ribosome.
- 28S, 5.8S, and 5S rRNA. One each of these molecules, along with some 45 different proteins, are used to make the large subunit of the ribosome.

The S number given each type of rRNA reflects the rate at which the molecules sediment in the ultracentrifuge. The larger the number, the larger the molecule (but not proportionally).

The 28S, 18S, and 5.8S molecules are produced by the processing of a single primary transcript from a cluster of identical copies of a single gene. The 5S molecules are produced from a different cluster of identical genes.

Transfer RNA (tRNA)

There are some 32 different kinds of tRNA in a typical eukaryotic cell.

- Each is the product of a separate gene.
- They are small (~4S), containing 73-93 nucleotides.

- Many of the bases in the chain pair with each other forming sections of double helix.
- The unpaired regions form 3 loops.
- Each kind of tRNA carries (at its 32 end) one of the 20 amino acids (thus most amino acids have more than one tRNA responsible for them).
- At one loop, 3 unpaired bases form an anticodon.
- Base pairing between the anticodon and the complementary codon on a mRNA molecule brings the correct amino acid into the growing polypeptide chain.

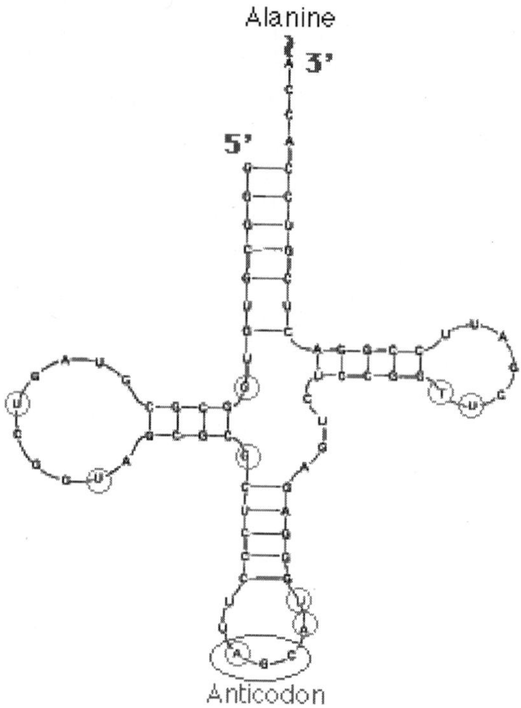

Fig. 15:

The four helical regions formed by base pairing are shown. The loop on the left binds to the enzyme that catalyzes the attachment (by means of a "high-energy" bond) of alanine to the molecule (top). The anticodon loop carries the three nucleotides that recognize the appropriate codon on the messenger RNA.

The circled letters indicate that the actual nucleotide is a chemically-modified form of the one shown.

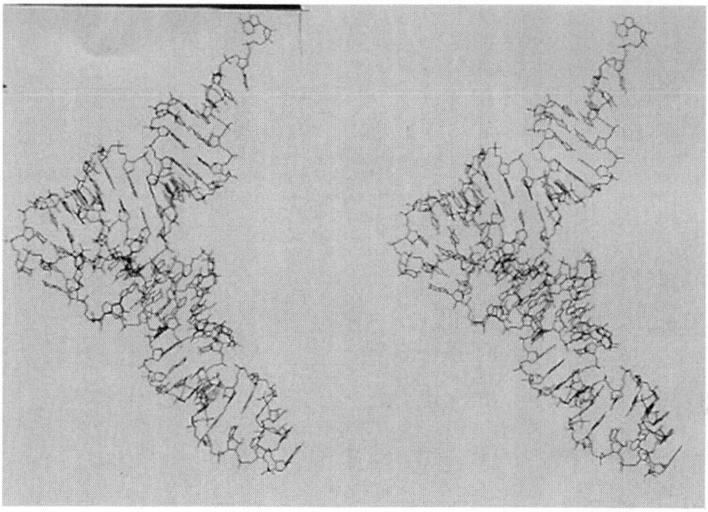

Fig. 16:

Above stereo view of phenylalanine transfer RNA from yeast. The 3' end where the phenylalanine is attached is at the upper right. The anticodon is located at the lower right. To fuse the two images, erect a stiff sheet of paper or cardboard between the two views so that your left eye sees only the left image and your right eye only the right image. With continued practice, you may even find that you can fuse the images with no props at all! (Courtesy of Dr. Sung-Hou Kim from J. L. Sussman and S-H. Kim, Science 192:853-858, May 28, 1976.)

Small Nuclear RNA (snRNA)

Approximately a dozen different genes for snRNAs, each present in multiple copies, have been identified. The snRNAs have various roles in the processing of the other classes of RNA. For example, several snRNAs are part of the spliceosomes that participate in converting pre-mRNA into mRNA by excising the introns and splicing the exons. [Link down to the discussion of RNA processing.]

Small Nucleolar RNA (snoRNA)

As the name suggests, these small (60–300 nucleotides) RNAs are found in the nucleolus where they are responsible for several functions:

- Some participate in making ribosomes by helping to cut up the large RNA precursor of the 28S, 18S, and 5.8S molecules.
- Others chemically modify many of the nucleotides in rRNA, tRNA, and snRNA molecules, e.g., by adding methyl groups to ribose.

- Some have been implicated in the alternative splicing of pre-mRNA to different forms of mature mRNA.
- One snoRNA serves as the template for the synthesis of telomeres.

In vertebrates, the snoRNAs are made from introns removed during RNA processing.

Noncoding RNA

Only messenger RNA encodes polypeptides. All the other classes of RNA, including types not mentioned here, are thus called noncoding RNA. Much remains to be learned about the function(s) of some of them. But, taken together, noncoding RNAs probably account for two-thirds of the transcription going on in the nucleus.

The RNA Polymerases

The RNA polymerases are huge multi-subunit protein complexes. Three kinds are found in eukaryotes.

- RNA polymerase I (Pol I). It transcribes the rRNA genes for the precursor of the 28S, 18S, and 5.8S molecules (and is the busiest of the RNA polymerases).
- RNA polymerase II (Pol II; also known as RNAP II). It transcribes protein-encoding genes into mRNA (and also the snRNA genes).
- RNA polymerase III (Pol III). It transcribes the 5S rRNA genes and all the tRNA genes.

Two remarkable reports in the 8 June 2001 issue of Science show the structure of Pol II and reveal many details of how it uses the antisense strand of DNA to synthesize a strand of mRNA.

RNA Processing: pre-mRNA '! mRNA

All the primary transcripts produced in the nucleus must undergo processing steps to produce functional RNA molecules for export to the cytosol. We shall confine ourselves to a view of the steps as they occur in the processing of pre-mRNA to mRNA.

Most eukaryotic genes are split into segments. In decoding the open reading frame of a gene for a known protein, one usually encounters periodic stretches of DNA calling for amino acids that do not occur in the actual protein product of that gene. Such stretches of DNA, which get transcribed into RNA but not translated into protein, are called introns. Those stretches of DNA that do code for amino acids in the protein are calledexons. Examples:

- The gene for one type of collagen found in chickens is split into 52 separate exons.
- The gene for dystrophin, which is mutated in boys with muscular dystrophy, has 79 exons.
- Even the genes for rRNA and tRNA are split by introns.
- The human genome is estimated to contain some 180,000 exons. With a current estimate of 21,000 genes, the average exon content of our genes is about 9.

In general, introns tend to be much longer than exons. An average eukaryotic exon is only 140 nucleotides long, but one human intron stretches for 480,000 nucleotides!

Removal of the introns — and splicing the exons together — are among the essential steps in synthesizing mRNA.

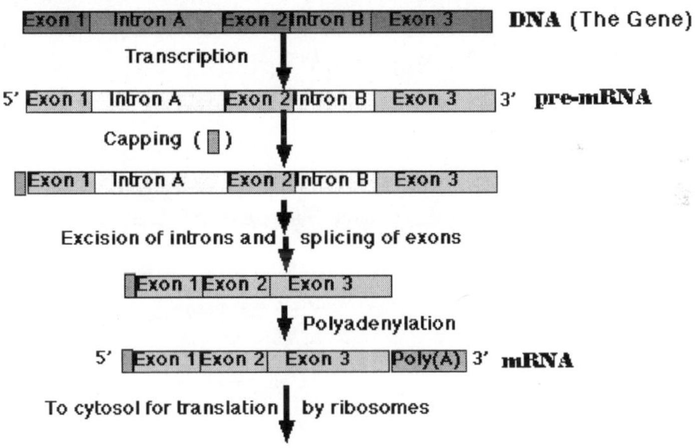

Fig. 17:

The steps of RNA processing:

- Synthesis of the cap. This is a modified guanine (G) which is attached to the 52 end of the pre-mRNA as it emerges from RNA polymerase II (RNAP II). The cap
  - o protects the RNA from being degraded by enzymes that degrade RNA from the 52 end;
  - o serves as an assembly point for the proteins needed to recruit the small subunit of the ribosome to begin translation.

- Step-by-step removal of introns present in the pre-mRNA and splicing of the remaining exons. This step takes place as the pre-mRNA continues to emerge from RNAP II.
- Synthesis of the poly(A) tail. This is a stretch of adenine (A) nucleotides. When a special poly(A) attachment site in the pre-mRNA emerges from RNAP II, the transcript is cut there, and the poly(A) tail is attached to the exposed 32 end. This completes the mRNA molecule, which is now ready for export to the cytosol. (The remainder of the transcript is degraded, and the RNA polymerase leaves the DNA.)

The cutting and splicing of mRNA must be done with great precision. If even one nucleotide is left over from an intron or one is removed from an exon, the reading frame from that point on will be shifted, producing new codons specifying a totally different sequence of amino acids from that point to the end of the molecule (which often ends prematurely anyway when the shifted reading frame generates a STOP codon).

The removal of introns and splicing of exons is done by spliceosomes. These are a complexes of 5 snRNA molecules and some 145 different proteins.

The introns in most pre-mRNAs begin with a GU and end with an AG. Presumably these short sequences assist in guiding the spliceosome.

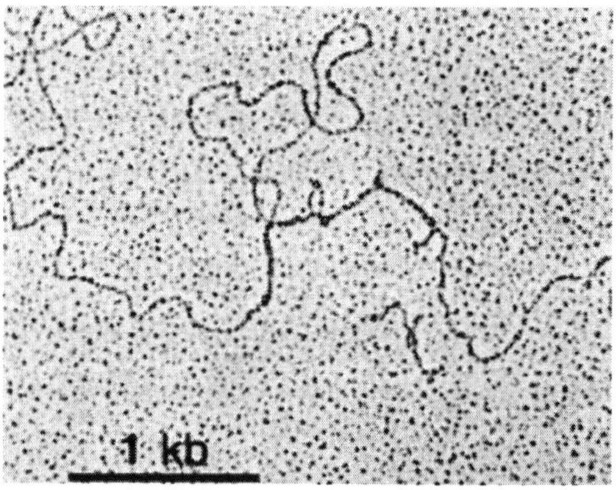

Fig. 18

Visual Evidence

The upper image is an electron micrograph of a mRNA-DNA hybrid molecule formed by mixing the messenger RNA (mRNA) from a clone of antibody-secreting cells with single-stranded DNA from the same kind of cells. The bar represents the length of 1000 bases.

The lower diagram is an interpretation of the micrograph. The solid line represents the DNA; the dotted line the mRNA. The loops (I_A, I_B, etc.) represent the introns that separate the exons encoding the domains of an antibody heavy chain:

- V = variable region
- E_1 = first constant region (C_H1) domain
- E_H = hinge region
- E_2 and E_3 = the nucleotides encoding the two C-terminal domains (C_H2 and C_H3).

[View a schematic of these domains.]

The unhybridized portion of the mRNA is its poly(A) tail.

[From R. Maki et al., Proc. Natl. Acad. Sci. USA 77:2138, 1980.]

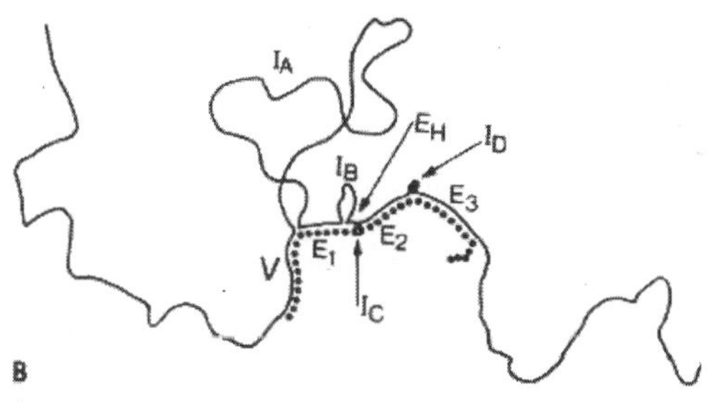

Fig. 19:

Alternative Splicing

The processing of pre-mRNA for many proteins proceeds along various paths in different cells or under different conditions. For example, early in the differentiation of a B cell (a lymphocyte that synthesizes an antibody) the cell first uses an exon that encodes a transmembrane domain that causes the molecule to be retained at the cell surface. Later, the B cell switches to using a different exon whose domain enables the protein to be secreted from the cell as a circulating antibody molecule.

Alternative splicing provides a mechanism for producing a wide variety of proteins from a small number of genes. While we humans may turn out to have only some 20 thousand genes, we probably make at least 10 times that number of different proteins. It is now estimated that 92–94% of our genes produce pre-mRNAs that are alternatively-spliced. There is evidence that the pattern of alternative splicing differs consistently in different tissuesand so must be regulated. But whether all the products are functional or that many are simply the outcome of an error-prone process remains to be seen.

Alternative splicing not only provides different proteins from a single gene but also different 3' UTRs and 5' UTRs. Although not translated into protein, these untranslated regions contain signals that, for example, dictate where in the cell that protein will accumulate. Two examples:

- The 3' UTR of the bicoid gene in Drosophila directs the mRNA to the anterior of the embryo [Link];
- the same region in the VegT gene of Xenopus directs its mRNA to the vegetal pole of the embryo [Link].

One of the most dramatic examples of alternative splicing is the Dscam gene in Drosophila. This single gene contains some 116 exons of which 17 are retained in the final mRNA. Some exons are always included; others are selected from an array. Theoretically this system is able to produce 38,016 different proteins. And, in fact, over 18,000 different ones have been found in Drosophila hemolymph.

These Dscam proteins are used to establish a unique identity for each neuron. It works like this. Each developing neuron synthesizes a dozen or so Dscam mRNAs out of the thousands of possibilities. Which ones are selected appears to be simply a matter of chance, but because of the great number of possibilities, each neuron will most likely end up with a unique set of a dozen or so Dscam proteins. As each developing neuron in the central nervous system sprouts dendrites and an axon, these express its unique collection of Dscam proteins. If the various extensions of a single neuron should meet each other in the tangled web that is the hallmark of nervous tissue, they are repelled. In this way, thousands of different neurons can coexist in intimate contact without the danger of nonfunctional contacts between the various extensions of the same neuron.

We also have a DSCAM gene which is located on chromosome 21. When present in 3 copies, it is probably responsible for some of the features of Down syndrome (and accounts for its full name: Down Syndrome CellAdhesion Molecule).

Perhaps the incredible diversity of synaptic junctions in the mammalian c.n.s. (~10) is mediated by alternative splicing of a limited number of gene transcripts.

So, whether a particular segment of RNA will be retained as an exon or excised as an intron can vary under different circumstances, such as

- what type of cell the gene is in;
- what stage of differentiation that cell is passing through;
- what extracellular signals that cell is receiving.

Clearly the switching to an alternate splicing pathway must be closely regulated.

Trans-splicing

Most genes are transcribed and their transcripts processed as described above. RNA polymerase travels down a single strand of a single gene locus to form pre-mRNA that is processed (including removal of introns) to form the mature mRNA. But there are exceptions. A number of cases have been found where two different precursor transcripts have been spliced together to form the final RNA molecule. The phenomenon is called trans-splicing.

Examples: synthesis of a single RNA molecule by splicing together transcripts from loci

- located far apart on the same chromosome or
- on opposite strands of the same gene locus or
- that are the two alleles of the gene on their separate (homologous) chromosomes.

The biological importance of these trans-spliced transcripts is still unknown for most of them.

Split Genes

Perhaps during evolution, eukaryotic genes have been assembled from smaller, primitive genes — today's exons. Some proteins, like the antibodies mentioned in the previous section, are organized in a set of separate sections or domains each with a special function to perform in the complete molecule. Each domain is encoded by a separate exon. Having the different functional parts of the antibody molecule encoded by separate exons makes it possible to use these units in different combinations. Thus a set of exons in the genome may be the genetic equivalent of the various modular pieces in a box of Lego® for children to assemble in whatever forms they wish.

But the boundaries of other exons do not seem to correspond to domain boundaries of the protein. So perhaps the major benefit of split genes is simply the opportunity they provide for making many different proteins from a single gene through alternative splicing.

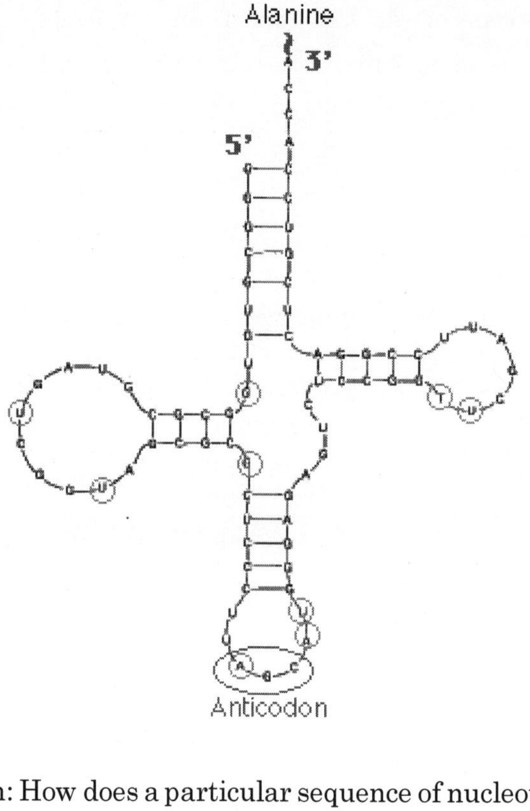

Fig. 20:

The problem: How does a particular sequence of nucleotides specify a particular sequence of amino acids?

The answer: by means of transfer RNA molecules, each specific for one amino acid and for a particular triplet of nucleotides in messenger RNA (mRNA) called acodon. The family of tRNA molecules enables the codons in a mRNA molecule to be translated into the sequence of amino acids in the protein.

This image shows the structure of alanine transfer RNA (tRNA) from yeast. It consists of a single strand of 77 ribonucleotides. The chain is folded on itself, and many of the bases pair with each other forming four helical regions. Loops are formed in the unpaired regions of the chain. (The bases circled in blue have been chemically-modified following synthesis of the molecule.)

At least one kind of tRNA is present for each of the 20 amino acids used in protein synthesis. (Some amino acids employ the services of two or three different tRNAs, so most cells contain as many as 32 different kinds of tRNA.) The amino acid is attached to the appropriate tRNA by an activating enzyme (one of 20 aminoacyl-tRNA synthetases) specific for that amino acid as well as for the tRNA assigned to it.

Each kind of tRNA has a sequence of 3 unpaired nucleotides — the anticodon — which can bind, following the rules of base pairing, to the complementary triplet of nucleotides — the codon — in a messenger RNA (mRNA) molecule. Just as DNA replication and transcription involve base pairing of nucleotides running in opposite direction, so the reading of codons in mRNA (5' -> 3') requires that the anticodons bind in the opposite direction.

Anticodon: 3' CGA 5'

Codon: 5' GCU 3'

The Steps of Translation

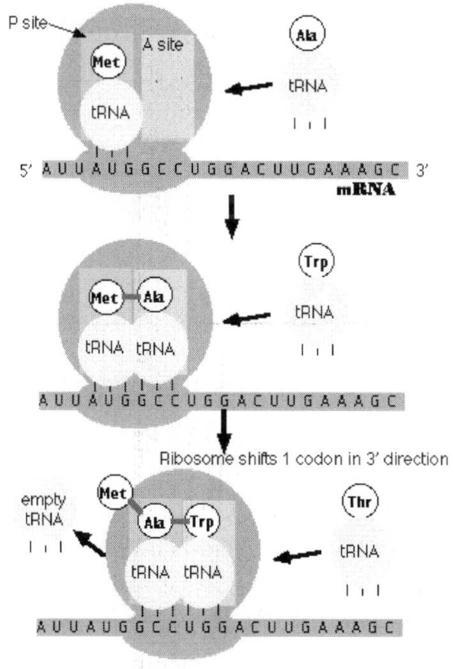

Fig. 21:

Initiation

- The small subunit of the ribosome binds to a site "upstream" (on the 5' side) of the start of the message.
- It proceeds downstream (5' -> 3') until it encounters the start codon AUG. (The region between the mRNA cap and the AUG is known as the 5'-untranslated region [5'-UTR].)
- Here it is joined by the large subunit and a special initiator tRNA.
- The initiator tRNA binds to the P site (shown in pink) on the ribosome.
- In eukaryotes, initiator tRNA carries methionine (Met). (Bacteria use a modified methionine designated fMet.)

Elongation

- An aminoacyl-tRNA (a tRNA covalently bound to its amino acid) able to base pair with the next codon on the mRNA arrives at the A site (green) associated with:
    - o an elongation factor (called EF-Tu in bacteria; EF-1 in eukaryotes)
    - o GTP (the source of the needed energy)
- The preceding amino acid (Met at the start of translation) is covalently linked to the incoming amino acid with a peptide bond (shown in red).
- The initiator tRNA is released from the P site.
- The ribosome moves one codon downstream.
- This shifts the more recently-arrived tRNA, with its attached peptide, to the P site and opens the A site for the arrival of a new aminoacyl-tRNA.
- This last step is promoted by another protein elongation factor (called EF-G in bacteria; EF-2 in eukaryotes) and the energy of another molecule of GTP.

Note: the initiator tRNA is the only member of the tRNA family that can bind directly to the P site. The P site is so-named because, with the exception of initiator tRNA, it binds only to a peptidyl-tRNA molecule; that is, a tRNA with the growing peptide attached.

The A site is so-named because it binds only to the incoming aminoacyl-tRNA; that is the tRNA bringing the next amino acid. So, for example, the tRNA that brings Met into the interior of the polypeptide can bind only to the A site.

Termination

- The end of translation occurs when the ribosome reaches one or more STOP codons (UAA, UAG, UGA). (The nucleotides from this point to the poly(A) tail make up the 3'-untranslated region [3'-UTR] of the mRNA.)
- There are no tRNA molecules with anticodons for STOP codons.
- However, protein release factors recognize these codons when they arrive at the A site.
- Binding of these proteins —along with a molecule of GTP — releases the polypeptide from the ribosome.
- The ribosome splits into its subunits, which can later be reassembled for another round of protein synthesis.

Polysomes

A single mRNA molecule usually has many ribosomes traveling along it, in various stages of synthesizing the protein. This complex is called a polysome.

Codon Bias

All but two of the amino acids (Met and Trp) can be encoded by from 2 to 6 different codons. However, the genome of most organisms reveals that certain codons are preferred over others. In humans, for example, alanine is encoded by GCC four times as often as by GCG. This probably reflects a greater translation efficiency by the translation apparatus for certain codons over their synonyms.

- At the start of translation, two or more of a set of synonymous codons (e.g., the 6 codons that incorporate leucine in the growing protein) are used alternately. The need to locate first one and then another tRNA for that amino acid slows down the rate of translation.
    - This may aid in keeping ribosomes from bumping into each other on the polysome.
    - It may also provide more time for the nascent protein to begin to fold correctly as it emerges from the ribosome.
- Once translation is well underway (after 30–50 amino acids have been added), one particular codon tends to be chosen each

time its amino acid is called for. Presumably this now increases the efficiency (speed) of translation.

- Most organisms have more than the 61 genes needed to encode a tRNA for each of the 61 codons (we have 270 tRNA genes). The presence of multiple genes for tRNAs with an identical anticodon increases the concentration of tRNAs able to bind a particular codon. Messenger RNAs — especially those of active genes — tend to favor codons that correspond to abundant tRNAs carrying the anticodon.

Codon bias even extends to pairs of codons: wherever a human protein contains the amino acids Ala-Glu, the gene encoding those amino acids is seven times as likely to use the codons GCAGAG rather than the synonymous GCCGAA.

Codon bias is exploited by the biotechnology industry to improve the yield of the desired product. The ability to manipulate codon bias may also usher in a era of safer vaccines.

Quality Control

Defective mRNA molecules can be produced by

- mutations in the gene as well as
- errors introduced during transcription (albeit at a remarkably low rate).

  In addition to producing mRNAs with incorrect codons for amino acids, these errors can produce mRNA molecules that have

- Premature Termination Codons (PTCs); that is, the introduction of a STOP codon before the normal end of the message. Translation of these mRNAs produces a truncated protein that is probably ineffective and may be harmful. The problem can sometimes be solved by Nonsense-Mediated mRNA Decay (NMD).
- no STOP codon. These produce "nonstop" transcripts. The problem can be solved by Nonstop mRNA Decay.

Nonsense-Mediated mRNA Decay (NMD)

Premature termination codons (PTCs) may be generated by

- "nonsense" mutations;
- frameshifts;

- RNA processing (intron removal) errors;
- as an inevitable consequence of creating antigen receptors on B cells and T cells.

Mechanism

- During RNA processing within the nucleus, protein complexes are added at each spot where adjacent exons are spliced together. (These are important signals for exporting the mRNA to the cytoplasm.)
- In the cytoplasm, as the ribosome moves down the mRNA, these complexes are removed (and sent back to the nucleus for reuse).
- If the ribosome encounters a premature termination codon, the final exon-exon tag(s) are not removed, and this marks the defective mRNA for destruction (in P bodies).

Mutations that introduce premature termination codons are responsible for some cases of such inherited human diseases as cystic fibrosis and Duchenne muscular dystrophy (DMD).

A drug, designated PTC124 or ataluren, causes the ribosome to skip over PTCs while still enabling normal termination of translation. PTC124 has shown promise in animal models of cystic fibrosis and DMD and phase II clinical trials are now being conducted on humans.

Nonstop mRNA Decay

Nonstop transcripts occur when there is no STOP codon in the message. As a result the ribosome is unable to recruit the release factors needed to leave the mRNA.

Nonstop transcripts are formed during RNA processing, e.g., by having the poly(A) tail put on before the STOP codon is reached.

Mechanisms

Eukaryotes and bacteria handle the problem of no STOP codon differently.

- In eukaryotes, when the ribosome stalls at the end of the poly(A) tail, proteins are recruited to
    - o release the ribosome for reuse and to
    - o degrade the faulty message.

- In bacteria, a special RNA molecule — called tmRNA saves the day. It is called tmRNA because it has the properties of both a transfer RNA and a messenger RNA.
  - o The transfer part adds alanine to the A site on the ribosome.
  - o The ribosome then moves on to the messenger part which encodes 10 amino acids that target the molecule for destruction (and releases the ribosome for reuse).

Regulation of Translation

The expression of most genes is controlled at the level of their transcription. Transcription factors (proteins) bind to promoters and enhancers turning on (or off) the genes they control.

However, gene expression can also be controlled at the level of translation.

By General RNA-Degradation Machinery

The cytosol of eukaryotes contains protein complexes that compete with ribosomes for access to mRNAs. As these increase their activity, they sequester mRNAs in larger aggregates called P bodies (for "processing bodies", but this processing should not be confused with the processing of pre-mRNA to mature mRNA that occurs in the nucleus).

These protein complexes break down the mRNA by

- removing its "cap";
- removing its poly(A) tail;
- degrading the remaining message (nibbling away in the 5' -> 3' direction).

What controls the dynamic balance between ribosomes and P bodies for access to mRNAs remains to be learned. But this mechanism provides for

- destruction of "bad" mRNAs (e.g., those with premature STOP codons.
- turnover of mRNAs thus increasing the flexibility of gene expression in the cell.

Exosomes

These are hollow macromolecular complexes with two openings. They take in unfolded RNA molecules and degrade them in the 3' -> 5' direction.

(In neither structure nor function do these exosomes resemble the exosomes involved in antigen presentation that unfortunately share the same name.)

By MicroRNAs (miRNAs)

Here small RNA molecules bind to a complementary portion in the 3'-UTR of the mRNA and

- prevent it from being translated by ribosomes and/or
- trigger its destruction.
- Both these activities take place in P bodies.

By Riboswitches

It turns out that the regulation of the level of certain metabolites is controlled by riboswitches. A riboswitch is a part of a molecule of messenger RNA (mRNA) with a specific binding site for the metabolite (or a close relative).

Examples:

- If thiamine pyrophosphate (the active form of thiamine [vitamin B_1]) is available in the culture medium of E. coli,
  - o it binds to a messenger RNA whose protein product is an enzyme needed to synthesize thiamine from the ingredients in minimal medium.
  - o Binding induces an allosteric shift in the structure of the mRNA so that it can no longer bind to a ribosome and thus cannot be translated into the enzyme.
  - o E. coli no longer wastes resources on synthesizing a vitamin that is available preformed.

A thiamine pyrophosphate riboswitch has also been found in plants, archaea, and Neurospora. The one in Neurospora regulates genes involved in vitamin B_1 metabolism by alternative splicing of their transcripts. (Other riboswitches act on transcription rather than translation [Link]).

- If vitamin B_{12} is present in the cell,
  - o it binds to the mRNA which encodes a protein needed to import the vitamin from the culture medium.
  - o This, too, induces an allosteric shift in the mRNA that prevents it from binding a ribosome.

- o E. coli no longer wastes resources on synthesizing a transporter for a vitamin that it already has enough of.
- Some Gram-positive bacteria (E. coli is Gram-negative) control the level of a sugar needed to synthesize their cell wall with a riboswitch. In this case, as the concentration of the sugar builds up, it binds to the messenger RNA (mRNA) whose product is the enzyme that makes the sugar. This causes the mRNA to self-destruct so production of the enzyme — and thus the sugar — ceases.

It has been suggested that these regulatory mechanisms, which do not involve any protein, are a relict from an "RNA world".

By Gene-Specific Proteins

Translation of at least one mRNA in humans is repressed by a protein — an aminoacyl tRNA synthetase. In response to the inflammatory cytokine interferon-gamma [IFN-ã], the synthetase abandons its normal function (adding Glu and Pro to their respective tRNAs) and instead binds to the mRNA blocking its translation.

In some bacteria, a protein product may inhibit the further translation of its own mRNA (a kind of feedback inhibition). It does so by binding to a site which blocks the mRNA from further association with a ribosome.

DNA Replication

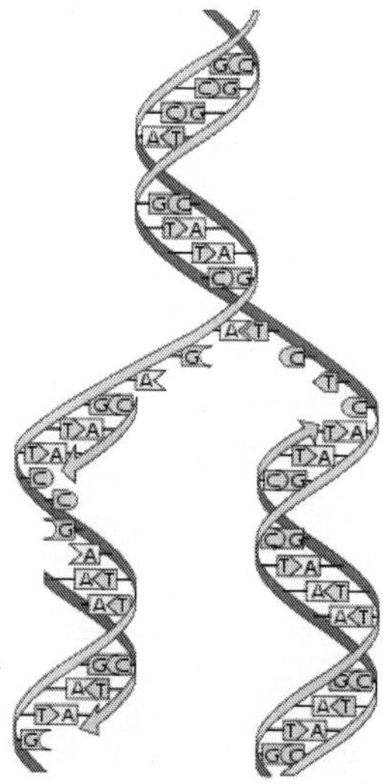

Fig. 21:

DNA replication. The double helix is unwound and each strand acts as a template for the next strand. Bases are matched to synthesize the new partner strands.

DNA replication is a biological process that occurs in all living organisms and copies their DNA; it is the basis for biological inheritance. The process starts with one double-stranded DNA molecule and produces two identical copies of the molecule. Each strand of the original double-stranded DNA molecule serves as template for the production of the complementary strand. Cellular proofreading and error toe-checking mechanisms ensure near perfect fidelity for DNA replication.

In a cell, DNA replication begins at specific locations in the genome, called "origins". Unwinding of DNA at the origin, and synthesis of new strands, forms areplication fork. In addition to DNA polymerase, the enzyme that synthesizes the new DNA by adding nucleotides matched to the template strand, a number of other proteins are associated with the fork and assist in the initiation and continuation of DNA synthesis.

DNA replication can also be performed in vitro (artificially, outside a cell). DNA polymerases, isolated from cells, and artificial DNA primers are used to initiate DNA synthesis at known sequences in a template molecule. The polymerase chain reaction (PCR), a common laboratory technique, employs such artificial synthesis in a cyclic manner to amplify a specific target DNA fragment from a pool of DNA.

Chapter 2

DNA Structure

The Chemical Structure of DNA

DNA usually exists as a double-stranded structure, with both strands coiled together to form the characteristic double-helix. Each single strand of DNA is a chain of four types of nucleotides having the bases: adenine, cytosine, guanine, and thymine.

A nucleotide is a mono-, di-, or triphosphate deoxyribonucleoside; that is, a deoxyribose sugar is attached to one, two, or three phosphates. Chemical interaction of these nucleotides forms phosphodiester linkages, creating the phosphate-deoxyribose backbone of the DNA double helix with the bases pointing inward. Nucleotides (bases) are matched between strands through hydrogen bonds to formbase pairs. Adenine pairs with thymine, and cytosine pairs with guanine.

DNA strands have a directionality, and the different ends of a single strand are called the "3' (three-prime) end" and the "5' (five-prime) end" with the direction of the naming going 5 prime to the 3 prime region.The strands of the helix are anti-parallel with one being 5 prime to 3 then the opposite strand 3 prime to 5 . These terms refer to the carbon atom in deoxyribose to which the next phosphate in the chain attaches.

In addition to being complementary, the two strands of DNA are antiparallel: They are orientated in opposite directions. This directionality has consequences in DNA synthesis, because DNA polymerase can synthesize DNA in only one direction by adding nucleotides to the 3' end of a DNA strand.

The pairing of bases in DNA through hydrogen bonding means that the information contained within each strand is redundant. The nucleotides on a single strand can be used to reconstruct nucleotides on a newly synthesized partner strand.

DNA Polymerase

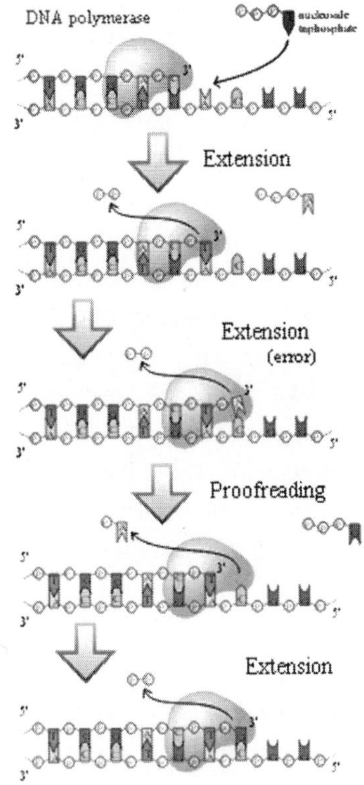

Fig. 1:

DNA polymerases adds nucleotides to the 5' end of a strand of DNA. If a mismatch is accidentally incorporated, the polymerase is inhibited from further extension. Proofreading removes the mismatched nucleotide and extension continues.

DNA polymerases are a family of enzymes that carry out all forms of DNA replication. However, a DNA polymerase can only extend an existing DNA strand paired with a template strand; it cannot begin the synthesis of a new strand. To begin synthesis, a short fragment of DNA or RNA, called a primer, must be created and paired with the template DNA strand.

DNA polymerase then synthesizes a new strand of DNA by extending the 3' end of an existing nucleotide chain, adding new nucleotides matched to the template strand one at a time via the creation of phosphodiester bonds. The energy for this process of DNA polymerization comes from two of the three total phosphates attached to each unincorporated base. (Free bases with their attached phosphate groups are called nucleoside triphosphates.)

When a nucleotide is being added to a growing DNA strand, two of the phosphates are removed and the energy produced creates a phosphodiester bond that attaches the remaining phosphate to the growing chain. The energetics of this process also help explain the directionality of synthesis - if DNA were synthesized in the 3' to 5' direction, the energy for the process would come from the 5' end of the growing strand rather than from free nucleotides.

In general, DNA polymerases are extremely accurate, making less than one mistake for every 10 nucleotides added. Even so, some DNA polymerases also have proofreading ability; they can remove nucleotides from the end of a strand in order to correct mismatched bases. If the 5' nucleotide needs to be removed during proofreading, the triphosphate end is lost. Hence, the energy source that usually provides energy to add a new nucleotide is also lost.

Replication Process

Origins

For a cell to divide, it must first replicate its DNA. This process is initiated at particular points in the DNA, known as "origins", which are targeted by proteins that separate the two strands and initiate DNA synthesis. Origins contain DNA sequences recognized by replication initiator proteins (e.g.,dnaA in E. coli' and the Origin Recognition Complex in yeast).

These initiator proteins recruit other proteins to separate the two strands and initiate replication forks.

Initiator proteins recruit other proteins to separate the DNA strands at the origin, forming a bubble.

Origins tend to be "AT-rich" (rich in adenine and thymine bases) to assist this process, because A-T base pairs have two hydrogen bonds (rather than the three formed in a C-G pair)—in general, strands rich in these nucleotides are easier to separate because a greater number of hydrogen bonds requires more energy to break them.

Once strands are separated, RNA primers are created on the template strands. To be more specific, the leading strand receives one RNA primer per active origin of replication while the lagging strand receives several; these several fragments of RNA primers found on the lagging strand of DNA are calledOkazaki fragments, named after their discoverer. DNA Polymerase extends the leading strand in one continuous motion and the lagging strand in a discontinuous motion (due to the Okazaki fragments). RNase removes the RNA fragments used to initiate replication by DNA Polymerase, and another DNA Polymerase enters to fill the gaps. When this is complete, a single nick on the leading strand and several nicks on the lagging strand can be found. Ligase works to fill these nicks in, thus completing the newly replicated DNA molecule.

As DNA synthesis continues, the original DNA strands continue to unwind on each side of the bubble, forming a replication fork with two prongs. In bacteria, which have a single origin of replication on their circular chromosome, this process eventually creates a "theta structure" (resembling the Greek letter theta: è). In contrast, eukaryotes have longer linear chromosomes and initiate replication at multiple origins within these.

Replication Fork

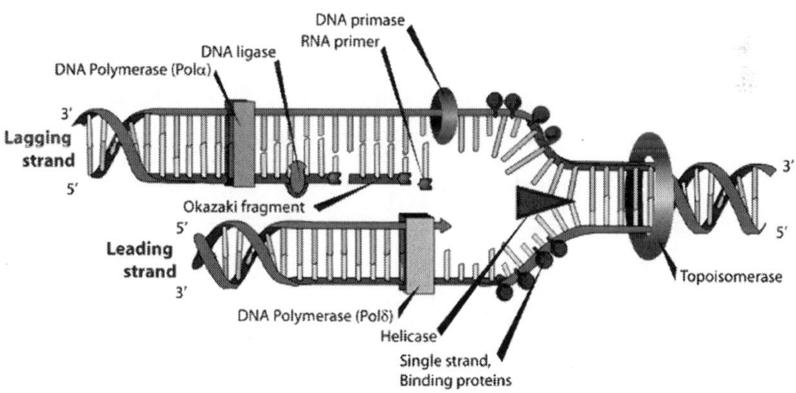

Fig. 2: Many enzymes are involved in the DNA replication fork.

The replication fork is a structure that forms within the nucleus during DNA replication. It is created by helicases, which break the hydrogen bonds holding the two DNA strands together. The resulting structure has two branching "prongs", each one made up of a single strand of DNA.

These two strands serve as the template for the leading and lagging strands, which will be created as DNA polymerase matches complementary nucleotides to the templates; The templates may be properly referred to as the leading strand template and the lagging strand template.

Leading Strand

The leading strand is the template strand of the DNA double helix so that the replication fork moves along it in the 3' to 5' direction. This allows the newly synthesized strand complementary to the original strand to be synthesized 5' to 3' in the same direction as the movement of the replication fork.

On the leading strand, a polymerase "reads" the DNA and adds nucleotides to it continuously. This polymerase is DNA polymerase III (DNA Pol III) in prokaryotes and presumably Pol å in yeasts.

In human cells the leading and lagging strands are synthesized by Pol á and Pol ä within the nucleus and Pol ã in the mitochondria. Pol å can substitute for Pol ä in special circumstances.

Lagging Strand

The lagging strand is the strand of the template DNA double helix that is oriented so that the replication fork moves along it in a 5' to 3' manner.

Because of its orientation, opposite to the working orientation of DNA polymerase III, which moves on a template in a 3' to 5' manner, replication of the lagging strand is more complicated than that of the leading strand.

On the lagging strand, primase "reads" the DNA and adds RNA to it in short, separated segments. In eukaryotes, primase is intrinsic to Pol á.

DNA polymerase III or Pol ä lengthens the primed segments, forming Okazaki fragments. Primer removal in eukaryotes is also performed by Pol ä.

In prokaryotes, DNA polymerase I "reads" the fragments, removes the RNA using its flap endonuclease domain (RNA primers are removed by 5'-3' exonuclease activity of polymerase I [weaver, 2005], and replaces the RNA nucleotides with DNA nucleotides (this is necessary because RNA and DNA use slightly different kinds of nucleotides). DNA ligase joins the fragments together.

Dynamics at the Replication Fork

Fig. 3: The assembled human DNA clamp, atrimer of the protein PCNA.

As helicase unwinds DNA at the replication fork, the DNA ahead is forced to rotate. This process results in a build-up of twists in the DNA ahead. This build-up would form a resistance that would eventually halt the progress of the replication fork. DNA Gyrase is an enzyme that temporarily breaks the strands of DNA, relieving the tension caused by unwinding the two strands of the DNA helix; DNA Gyrase achieves this by adding negative supercoils to the DNA helix.

Bare single-stranded DNA tends to fold back on itself and form secondary structures; these structures can interfere with the movement of DNA polymerase. To prevent this, single-strand binding proteins bind to the DNA until a second strand is synthesized, preventing secondary structure formation.

Clamp proteins form a sliding clamp around DNA, helping the DNA polymerase maintain contact with its template, thereby assisting with processivity. The inner face of the clamp enables DNA to be threaded through it. Once the polymerase reaches the end of the template or detects double-stranded DNA, the sliding clamp undergoes a conformational change that releases the DNA polymerase. Clamp-loading proteins are used to initially load the clamp, recognizing the junction between template and RNA primers.

Regulation

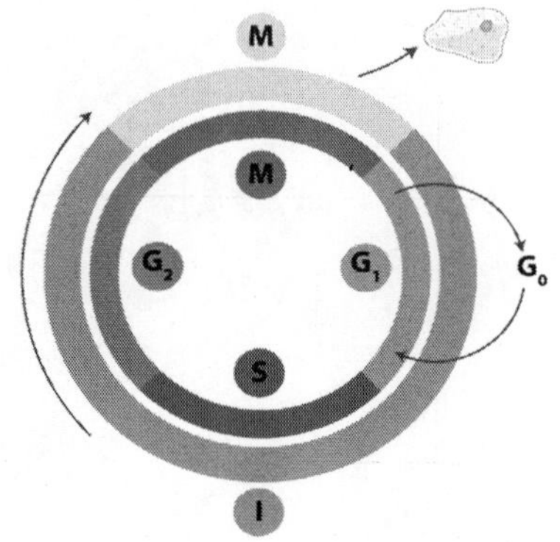

Fig. 4: The cell cycle of eukaryotic cells.

Eukaryotes

Within eukaryotes, DNA replication is controlled within the context of the cell cycle. As the cell grows and divides, it progresses through stages in the cell cycle; DNA replication occurs during the S phase (synthesis phase). The progress of the eukaryotic cell through the cycle is controlled by cell cycle checkpoints. Progression through checkpoints is controlled through complex interactions between various proteins, including cyclins and cyclin-dependent kinases.

The G1/S checkpoint (or restriction checkpoint) regulates whether eukaryotic cells enter the process of DNA replication and subsequent division. Cells that do not proceed through this checkpoint are remain in the G0 stage and do not replicate their DNA.

Replication of chloroplast and mitochondrial genomes occurs independent of the cell cycle, through the process of D-loop replication.

Bacteria

Most bacteria do not go through a well-defined cell cycle but instead continuously copy their DNA; during rapid growth, this can result in the concurrent occurrences of multiple rounds of replication. In E. coli, the best-characterized bacteria, DNA replication is regulated through several mechanisms, including: the hemimethylation and sequestering of the origin sequence, the ratio of ATP to ADP, and the levels of protein DnaA. All these control the process of initiator proteins binding to the origin sequences.

Because E. coli methylates GATC DNA sequences, DNA synthesis results in hemimethylated sequences. This hemimethylated DNA is recognized by the protein SeqA, which binds and sequesters the origin sequence; in addition, dnaA (required for initiation of replication) binds less well to hemimethylated DNA. As a result, newly replicated origins are prevented from immediately initiating another round of DNA replication.

ATP builds up when the cell is in a rich medium, triggering DNA replication once the cell has reached a specific size. ATP competes with ADP to bind to DnaA, and the DnaA-ATP complex is able to initiate replication. A certain number of DnaA proteins are also required for DNA replication — each time the origin is copied, the number of binding sites for DnaA doubles, requiring the synthesis of more DnaA to enable another initiation of replication.

Termination

Eukaryotes initiate DNA replication at multiple points in the chromosome, so replication forks meet and terminate at many points in the chromosome; these are not known to be regulated in any particular way. Because eukaryotes have linear chromosomes, DNA replication is unable to reach the very end of the chromosomes, but ends at the telomere region of repetitive DNA close to the end. This shortens the telomere of the daughter DNA strand. This is a normal process in somatic cells. As a result, cells can only divide a certain number of times before the DNA loss prevents further division. (This is known as the Hayflick limit.) Within the germ cell line, which passes DNA to the next generation, telomerase extends the repetitive sequences of the telomere region to prevent degradation. Telomerase can become mistakenly active in somatic cells, sometimes leading to cancer formation.

Because bacteria have circular chromosomes, termination of replication occurs when the two replication forks meet each other on the opposite end of the parental chromosome. E coli regulate this process through the use of termination sequences that, when bound by the Tus protein, enable only one direction of replication fork to pass through. As a result, the replication forks are constrained to always meet within the termination region of the chromosome.

Polymerase Chain Reaction

Researchers commonly replicate DNA in vitro using the polymerase chain reaction (PCR). PCR uses a pair of primers to span a target region in template DNA, and then polymerizes partner strands in each direction from these primers using a thermostable DNA polymerase. Repeating this process through multiple cycles produces amplification of the targeted DNA region. At the start of each cycle, the mixture of template and primers is heated, separating the newly synthesized molecule and template. Then, as the mixture cools, both of these become templates for annealing of new primers, and the polymerase extends from these. As a result, the number of copies of the target region doubles each round, increasing exponentially.

Virus

A virus is a small infectious agent that can replicate only inside the living cells of organisms. Viruses infect all types of organisms, from animals and plants tobacteria and archaea. Since Dmitri Ivanovsky's 1892 article describing a non-bacterial pathogen infecting tobacco plants, and the discovery of the tobacco mosaic virus by Martinus Beijerinck in 1898, about 5,000 viruses have been described in detail, although there are millions of different types. Viruses are found in almost every ecosystem on Earth and are the most abundant type of biological entity. The study of viruses is known as virology, a sub-speciality of microbiology.

Virus particles (known as virions) consist of two or three parts: the genetic material made from either DNA or RNA, long molecules that carry genetic information; a protein coat that protects these genes; and in some cases an envelope of lipids that surrounds the protein coat when they are outside a cell. The shapes of viruses range from simple helical and icosahedral forms to more complex structures. The average virus is about one one-hundredth the size of the average bacterium. Most viruses are too small to be seen directly with a light microscope.

The origins of viruses in the evolutionary history of life are unclear: some may have evolved from plasmids – pieces of DNA that can move between cells – while others may have evolved from bacteria. In evolution, viruses are an important means of horizontal gene transfer, which increases genetic diversity.

Viruses spread in many ways; viruses in plants are often transmitted from plant to plant by insects that feed on the sap of plants, such as aphids; viruses inanimals can be carried by blood-sucking insects. These disease-bearing organisms are known as vectors. Influenza viruses are spread by coughing and sneezing. Norovirus and rotavirus, common causes of viral gastroenteritis, are transmitted by the faecal-oral route and are passed from person to person by contact, entering the body in food or water. HIV is one of several viruses transmitted through sexual contact and by exposure to infected blood. The range of host cells that a virus can infect is called its "host range". This can be narrow or, as when a virus is capable of infecting many species, broad.

Viral infections in animals provoke an immune response that usually eliminates the infecting virus. Immune responses can also be produced by vaccines, which confer an artificially acquired immunity to the specific viral infection. However, some viruses including those causing AIDS and viral hepatitis evade these immune responses and result in chronic infections. Antibiotics have no effect on viruses, but several antiviral drugs have been developed.

Etymology

The word is from the Latin virus referring to poison and other noxious substances, first used in English in 1392. Virulent, from Latin virulentus (poisonous), dates to 1400. A meaning of "agent that causes infectious disease" is first recorded in 1728, before the discovery of viruses by Dmitry Ivanovsky in 1892. The plural is viruses. The adjective viral dates to 1948. The term virion is also used to refer to a single infective viral particle.

History

Fig. 5: Martinus Beijerinck in his laboratory in 1921

Louis Pasteur was unable to find a causative agent for rabies and speculated about a pathogen too small to be detected using a microscope. In 1884, the French microbiologist Charles Chamberland invented a filter (known today as the Chamberland filter or Chamberland-Pasteur filter) with pores smaller than bacteria. Thus, he could pass a solution containing bacteria through the filter and completely remove them from the solution.

In 1892, the Russian biologistDmitry Ivanovsky used this filter to study what is now known as the tobacco mosaic virus. His experiments showed that crushed leaf extracts from infected tobacco plants remain infectious after filtration. Ivanovsky suggested the infection might be caused by a toxin produced by bacteria, but did not pursue the idea. At the time it was thought that all infectious agents could be retained by filters and grown on a nutrient medium – this was part of the germ theory of disease.

In 1898, the Dutch microbiologist Martinus Beijerinck repeated the experiments and became convinced that the filtered solution contained a new form of infectious agent. He observed that the agent multiplied only in cells that were dividing, but as his experiments did not show that it was made of particles, he called it a contagium vivum fluidum (soluble living germ) and re-introduced the word virus. Beijerinck maintained that viruses were liquid in nature, a theory later discredited by Wendell Stanley, who proved they were particulate. In the same year Friedrich Loeffler and Frosch passed the first animal virus – agent of foot-and-mouth disease (aphthovirus) – through a similar filter.

In the early 20th century, the English bacteriologist Frederick Twort discovered a group of viruses that infect bacteria, now called bacteriophages (or commonly phages), and the French-Canadian microbiologist Félix d'Herelle described viruses that, when added to bacteria on agar, would produce areas of dead bacteria.

He accurately diluted a suspension of these viruses and discovered that the highest dilutions (lowest virus concentrations), rather than killing all the bacteria, formed discrete areas of dead organisms. Counting these areas and multiplying by the dilution factor allowed him to calculate the number of viruses in the original suspension. Phages were heralded as a potential treatment for diseases such as typhoid and cholera, but their promise was forgotten with the development of penicillin. The study of phages provided insights into the switching on and off of genes, and a useful mechanism for introducing foreign genes into bacteria.

By the end of the 19th century, viruses were defined in terms of their infectivity, their ability to be filtered, and their requirement for living hosts. Viruses had been grown only in plants and animals. In 1906, Ross Granville Harrison invented a method for growing tissue in lymph, and, in 1913, E. Steinhardt, C. Israeli, and R. A. Lambert used this method to grow vaccinia virus in fragments of guinea pig corneal tissue. In 1928, H. B. Maitland and M. C. Maitland grew vaccinia virus in suspensions of minced hens' kidneys. Their method was not widely adopted until the 1950s, when poliovirus was grown on a large scale for vaccine production.

Another breakthrough came in 1931, when the American pathologist Ernest William Goodpasture grew influenza and several other viruses in fertilized chickens' eggs. In 1949, John F. Enders,Thomas Weller, and Frederick Robbins grew polio virus in cultured human embryo cells, the first virus to be grown without using solid animal tissue or eggs. This work enabled Jonas Salk to make an effective polio vaccine.

The first images of viruses were obtained upon the invention of electron microscopy in 1931 by the German engineers Ernst Ruska and Max Knoll. In 1935, American biochemist and virologistWendell Meredith Stanley examined the tobacco mosaic virus and found it was mostly made of protein. A short time later, this virus was separated into protein and RNA parts. The tobacco mosaic virus was the first to be crystallised and its structure could therefore be elucidated in detail. The first X-ray diffraction pictures of the crystallised virus were obtained by Bernal and Fankuchen in 1941. On the basis of her pictures, Rosalind Franklin discovered the full DNA structure of the virus in 1955. In the same year, Heinz Fraenkel-Conrat and Robley Williams showed that purified tobacco mosaic virus RNA and its coat protein can assemble by themselves to form functional viruses, suggesting that this simple mechanism was probably the means through which viruses were created within their host cells.

The second half of the 20th century was the golden age of virus discovery and most of the 2,000 recognised species of animal, plant, and bacterial viruses were discovered during these years. In 1957, equine arterivirus and the cause of Bovine virus diarrhea (a pestivirus) were discovered. In 1963, the hepatitis B virus was discovered by

Baruch Blumberg, and in 1965, Howard Temin described the first retrovirus. Reverse transcriptase, the key enzyme that retroviruses use to translate their RNA into DNA, was first described in 1970, independently by Howard Martin Temin and David Baltimore. In 1983 Luc Montagnier's team at the Pasteur Institute in France, first isolated the retrovirus now called HIV.

Origins

Viruses are found wherever there is life and have probably existed since living cells first evolved. The origin of viruses is unclear because they do not form fossils, so molecular techniques have been used to compare the DNA or RNA of viruses and are a useful means of investigating how they arose. There are three main hypotheses that try to explain the origins of viruses:

Regressive Hypothesis

Viruses may have once been small cells that parasitised larger cells. Over time, genes not required by their parasitism were lost. The bacteria rickettsia and chlamydia are living cells that, like viruses, can reproduce only inside host cells. They lend support to this hypothesis, as their dependence on parasitism is likely to have caused the loss of genes that enabled them to survive outside a cell. This is also called the degeneracy hypothesis, or reduction hypothesis.

Cellular Origin Hypothesis

Some viruses may have evolved from bits of DNA or RNA that "escaped" from the genes of a larger organism. The escaped DNA could have come from plasmids (pieces of naked DNA that can move between cells) or transposons (molecules of DNA that replicate and move around to different positions within the genes of the cell). Once called "jumping genes", transposons are examples of mobile genetic elements and could be the origin of some viruses. They were discovered in maize by Barbara McClintock in 1950. This is sometimes called the vagrancy hypothesis, or the escape hypothesis.

Coevolution Hypothesis

This is also called the virus-first hypothesis and proposes that viruses may have evolved from complex molecules of protein and nucleic acid at the same time as cells first appeared on earth and would have been dependent on cellular life for billions of years. Viroids are molecules of RNA that are not classified as viruses because they lack a protein coat. However, they have characteristics that are common to several viruses and are often called subviral agents. Viroids are important pathogens of plants. They do not code for proteins but interact with the host cell and use the host machinery for their replication. The hepatitis delta virus of humans has an RNA genome similar to viroids but has a protein coat derived from hepatitis B virus and cannot produce one of its own. It is, therefore, a defective virus and cannot replicate without the help of hepatitis B virus. In similar manner, the virophage 'sputnik' is dependent on mimivirus, which infects the protozoanAcanthamoeba castellanii. These viruses that are dependent on the presence of other virus species in the host cell are called satellites and may represent evolutionary intermediates of viroids and viruses.

In the past, there were problems with all of these hypotheses: the regressive hypothesis did not explain why even the smallest of cellular parasites do not resemble viruses in any way. The escape hypothesis did not explain the complex capsids and other structures on virus particles. The virus-first hypothesis contravened the definition of viruses in that they require host cells. Viruses are now recognised as ancient and to have origins that pre-date the divergence of life into the three domains. This discovery has led modern virologists to reconsider and re-evaluate these three classical hypotheses.

The evidence for an ancestral world of RNA cells and computer analysis of viral and host DNA sequences are giving a better understanding of the evolutionary relationships between different viruses and may help identify the ancestors of modern viruses. To date, such analyses have not proved which of these hypotheses is correct. However, it seems unlikely that all currently known viruses have a common ancestor, and viruses have probably arisen numerous times in the past by one or more mechanisms.

Prions are infectious protein molecules that do not contain DNA or RNA. They cause an infection in sheep called scrapie and cattle bovine spongiform encephalopathy ("mad cow" disease). In humans they cause kuru and Creutzfeldt-Jakob disease. They are able to replicate because some proteins can exist in two different shapes and the prion changes the normal shape of a host protein into the prion shape. This starts a chain reaction where each prion protein converts many host proteins into more prions, and these new prions then go on to convert even more protein into prions. Although they are fundamentally different from viruses and viroids, their discovery gives credence to the idea that viruses could have evolved from self-replicating molecules.

Microbiology

Life Properties

Opinions differ on whether viruses are a form of life, or organic structures that interact with living organisms. They have been described as "organisms at the edge of life", since they resemble organisms in that they possess genes and evolve by natural selection, and reproduce by creating multiple copies of themselves through self-assembly. Although they have genes, they do not have a cellular structure, which is often seen as the basic unit of life. Viruses do not have their own metabolism, and require a host cell to make new products. They therefore cannot naturally reproduce outside a host cell – although bacterial species such as rickettsia and chlamydia are considered living organisms despite the same limitation. Accepted forms of life use cell division to reproduce, whereas viruses spontaneously assemble within cells. They differ from autonomous growth of crystals as they inherit genetic mutations while being subject to natural selection. Virus self-assembly within host cells has implications for the study of the origin of life, as it lends further credence to the hypothesis that life could have started as self-assembling organic molecules.

Structure

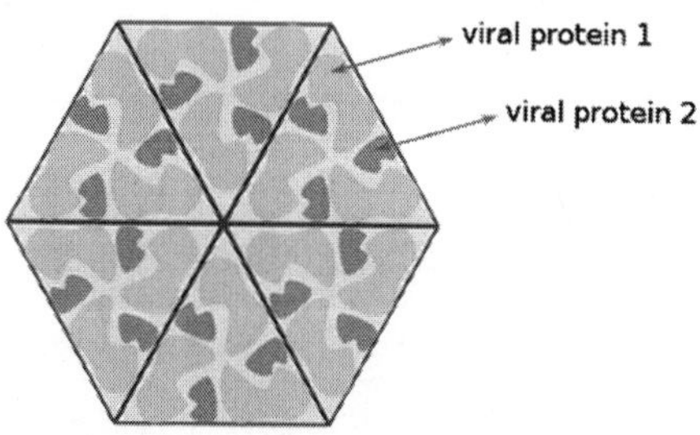

Hexon of a virus capsid made from two protein molecules

Fig. 6:

Diagram of how a virus capsid can be constructed using multiple copies of just twoprotein molecules

Viruses display a wide diversity of shapes and sizes, called morphologies. Generally viruses are much smaller than bacteria. Most viruses that have been studied have a diameter between 20 and 300 nanometres. Some filoviruses have a total length of up to 1400 nm; their diameters are only about 80 nm.Most viruses cannot be seen with a light microscope so scanning and transmission electron microscopes are used to visualise virions. To increase the contrast between viruses and the background, electron-dense "stains" are used. These are solutions of salts of heavy metals, such as tungsten, that scatter the electrons from regions covered with the stain. When virions are coated with stain (positive staining), fine detail is obscured. Negative staining overcomes this problem by staining the background only.

A complete virus particle, known as a virion, consists of nucleic acid surrounded by a protective coat of protein called a capsid. These are formed from identical protein subunits called capsomeres. Viruses can have a lipid "envelope" derived from the host cell membrane. The capsid is made from proteins encoded by the viral genome and its shape serves as the basis for morphological distinction. Virally coded protein subunits will self-assemble to form a capsid, generally requiring the presence of the virus genome. Complex viruses code for proteins that assist in the construction of their capsid. Proteins associated with nucleic acid are known as nucleoproteins, and the association of viral capsid proteins with viral nucleic acid is called a nucleocapsid. The capsid and entire virus structure can be mechanically (physically) probed through atomic force microscopy. In general, there are four main morphological virus types:

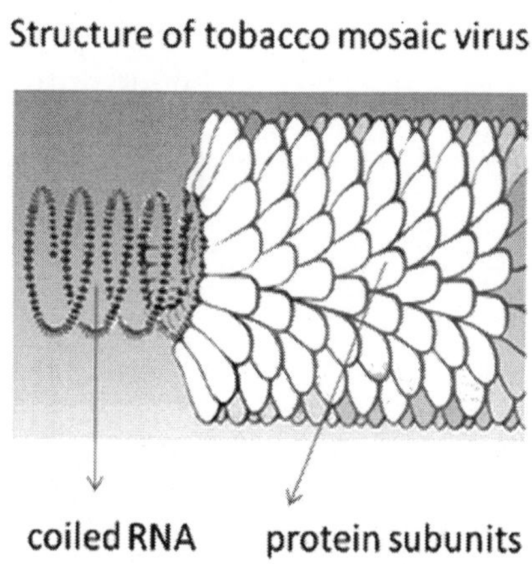

Fig. 7: RNA coiled in a helix of repeating protein sub-units

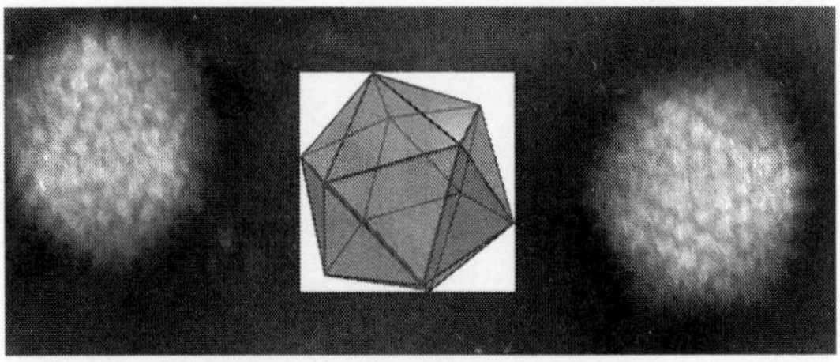

Fig. 8: Electron micrograph of icosahedraladenovirus

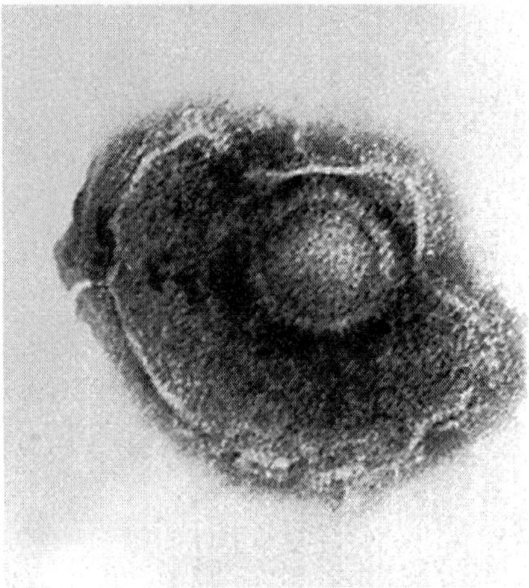

Fig. 9: Herpes viruses have a lipid envelope

Helical

These viruses are composed of a single type of capsomer stacked around a central axis to form a helical structure, which may have a central cavity, or hollow tube. This arrangement results in rod-shaped or filamentous virions: These can be short and highly rigid, or long and very flexible. The genetic material, in general, single-stranded RNA, but ssDNA in some cases, is bound into the protein helix by interactions between the negatively charged nucleic acid and positive charges on the protein. Overall, the length of a helical capsid is related to the length of the nucleic acid contained within it and the diameter is dependent on the size and arrangement of capsomers. The well-studied tobacco mosaic virus is an example of a helical virus.

Icosahedral

Most animal viruses are icosahedral or near-spherical with icosahedral symmetry. A regular icosahedron is the optimum way of forming a closed shell from identical sub-units. The minimum number of identical capsomers required is twelve, each composed of five identical sub-units. Many viruses, such as rotavirus, have more than twelve capsomers and appear spherical but they retain this symmetry. Capsomers at the apices are surrounded by five other capsomers and are called pentons. Capsomers on the triangular faces are surrounded by six others and are called hexons. Hexons are essentially flat and pentons, which form the 12 vertices, are curved. The same protein may act as the subunit of both the pentamers and hexamers or they may be composed of different proteins.

Although the isosahedral structure is extremely common among viruses because of size differences slight variations exist between virions. Given an asymmetric subunit on a triangular face of a regular isosahedron, with three subunits per face 60 such subunits can be placed in an equivalent manner. Most virions, because of their size, have more than 60 subunits. These variations have been classified on the basis of the quasi-equivalence principle proposed by Caspar and Klug.

An isosahedral structure can be regarded as being constructed from 12 pentamers. The number of pentamers is fixed but the number of hexamers can vary.These shells can be constructed from pentamers and hexamers by minimizing the number T (triangulation number) of nonequivalent locations that subunits occupy, with the T-number adopting the particular integer values 1, 3, 4, 7, 12, 13,...($T = h + k + hk$, with h, k equal to nonnegative integers). These shells always contain 12 pentamers plus 10 (T-1) hexamers. Although this classification can be applied to the majority of known viruses exceptions are known including the retroviruses where point mutations disrupt the symmetry.

Prolate

This is an isosahedron elongated along the fivefold axis and is a common arrangement of the heads of bacteriophages. Such a structure is composed of a cylinder with a cap at either end. The cylinder is composed of 10 triangles. The Q number, which can be can be any positive integer, specifies the number of triangles, composed of asymmetric subunits, that make up the 10 triangles of the cylinder. The caps are classified by the T number.

Envelope

Some species of virus envelop themselves in a modified form of one of the cell membranes, either the outer membrane surrounding an infected host cell or internal membranes such as nuclear membrane or endoplasmic reticulum, thus gaining an outer lipid bilayer known as a viral envelope.

This membrane is studded with proteins coded for by the viral genome and host genome; the lipid membrane itself and any carbohydrates present originate entirely from the host. The influenza virus and HIV use this strategy. Most enveloped viruses are dependent on the envelope for their infectivity.

Complex

These viruses possess a capsid that is neither purely helical nor purely icosahedral, and that may possess extra structures such as protein tails or a complex outer wall. Some bacteriophages, such as Enterobacteria phage T4, have a complex structure consisting of an icosahedral head bound to a helical tail, which may have a hexagonal base plate with protruding protein tail fibres. This tail structure acts like a molecular syringe, attaching to the bacterial host and then injecting the viral genome into the cell.

The poxviruses are large, complex viruses that have an unusual morphology. The viral genome is associated with proteins within a central disk structure known as a nucleoid. The nucleoid is surrounded by a membrane and two lateral bodies of unknown function. The virus has an outer envelope with a thick layer of protein studded over its surface.

The whole virion is slightly pleiomorphic, ranging from ovoid to brick shape. Mimivirus is the largest characterised virus, with a capsid diameter of 400 nm. Protein filaments measuring 100 nm project from the surface. The capsid appears hexagonal under an electron microscope, therefore the capsid is probably icosahedral. In 2011, researchers discovered a larger virus on ocean floor of the coast of Las Cruces, Chile. Provisionally namedMegavirus chilensis, it can be seen with a basic light microscope.

Some viruses that infect Archaea have complex structures that are unrelated to any other form of virus, with a wide variety of unusual shapes, ranging from spindle-shaped structures, to viruses that resemble hooked rods, teardrops or even bottles. Other archaeal viruses resemble the tailed bacteriophages, and can have multiple tail structures.

Genome

Table 1: Genomic Diversity Among Viruses

Property	*Parameters*
Nucleic acid	" DNA
	" RNA
	" Both DNA and RNA (at different stages in the life cycle)
Shape	" Linear
	" Circular
	" Segmented
Strandedness	" Single-stranded
	" Double-stranded
	" Double-stranded with regions of single-strandedness
Sense	
	" Positive sense (+)
	" Negative sense (?)
	" Ambisense (+/?)

An enormous variety of genomic structures can be seen among viral species; as a group they contain more structural genomic diversity than plants, animals, archaea, or bacteria. There are millions of different types of viruses, although only about 5,000 of them have been described in detail. A virus has either DNA or RNA genes and is called a DNA virus or a RNA virus respectively. The vast majority of viruses have RNA genomes. Plant viruses tend to have single-stranded RNA genomes and bacteriophages tend to have double-stranded DNA genomes.

Viral genomes are circular, as in the polyomaviruses, or linear, as in the adenoviruses. The type of nucleic acid is irrelevant to the shape of the genome. Among RNA viruses and certain DNA viruses, the genome is often divided up into separate parts, in which case it is called segmented. For RNA viruses, each segment often codes for only one protein and they are usually found together in one capsid. However, all segments are not required to be in the same virion for the virus to be infectious, as demonstrated by brome mosaic virus and several other plant viruses.

A viral genome, irrespective of nucleic acid type, is almost always either single-stranded or double-stranded. Single-stranded genomes consist of an unpaired nucleic acid, analogous to one-half of a ladder split down the middle. Double-stranded genomes consist of two complementary paired nucleic acids, analogous to a ladder. The virus particles of some virus families, such as those belonging to the Hepadnaviridae, contain a genome that is partially double-stranded and partially single-stranded.

For most viruses with RNA genomes and some with single-stranded DNA genomes, the single strands are said to be either positive-sense (called the plus-strand) or negative-sense (called the minus-strand), depending on whether or not they are complementary to the viral messenger RNA (mRNA). Positive-sense viral RNA is in the same sense as viral mRNA and thus at least a part of it can be immediately translated by the host cell. Negative-sense viral RNA is complementary to mRNA and thus must be converted to positive-sense RNA by an RNA-dependent RNA polymerase before translation. DNA nomenclature for viruses with single-sense genomic ssDNA is similar to RNA nomenclature, in that the coding strand for the viral mRNA is complementary to it ("), and the non-coding strand is a copy of it (+). However, several types of ssDNA and ssRNA viruses have genomes that are ambisense in that transcription can occur off both strands in a double-stranded replicative intermediate. Examples include geminiviruses, which are ssDNA plant viruses andarenaviruses, which are ssRNA viruses of animals.

Genome size varies greatly between species. The smallest viral genomes – the ssDNA circoviruses, family Circoviridae – code for only two proteins and have a genome size of only 2 kilobases; the largest – mimiviruses – have genome sizes of over 1.2 megabases and code for over one thousand proteins. RNA viruses generally have smaller genome sizes than DNA viruses because of a higher error-rate when replicating, and have a maximum upper size limit. Beyond this limit, errors in the genome when replicating render the virus useless or uncompetitive. To compensate for this, RNA viruses often have segmented genomes – the genome is split into smaller molecules – thus reducing the chance that an error in a single-component genome will incapacitate the entire genome. In contrast, DNA viruses generally have larger genomes because of the high fidelity of their replication enzymes. Single-strand DNA viruses are an exception to this rule, however, as mutation rates for these genomes can approach the extreme of the ssRNA virus case.

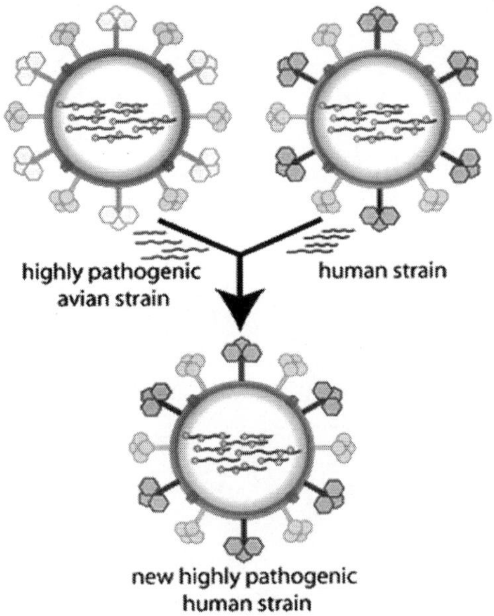

Fig. 10:

How antigenic shift, or reassortment, can result in novel and highly pathogenic strains of human influenza.

Viruses undergo genetic change by several mechanisms. These include a process called genetic drift where individual bases in the DNA or RNA mutate to other bases. Most of these point mutations are "silent" – they do not change the protein that the gene encodes – but others can confer evolutionary advantages such as resistance to antiviral drugs. Antigenic shift occurs when there is a major change in the genome of the virus. This can be a result ofrecombination or reassortment. When this happens with influenza viruses, pandemics might result. RNA viruses often exist as quasispecies or swarms of viruses of the same species but with slightly different genome nucleoside sequences. Such quasispecies are a prime target for natural selection.

Segmented genomes confer evolutionary advantages; different strains of a virus with a segmented genome can shuffle and combine genes and produce progeny viruses or (offspring) that have unique characteristics. This is called reassortment or viral sex.

Genetic recombination is the process by which a strand of DNA is broken and then joined to the end of a different DNA molecule. This can occur when viruses infect cells simultaneously and studies of viral evolution have shown that recombination has been rampant in the species studied. Recombination is common to both RNA and DNA viruses.

Replication cycle

Viral populations do not grow through cell division, because they are acellular. Instead, they use the machinery and metabolism of a host cell to produce multiple copies of themselves, and they assemble in the cell.

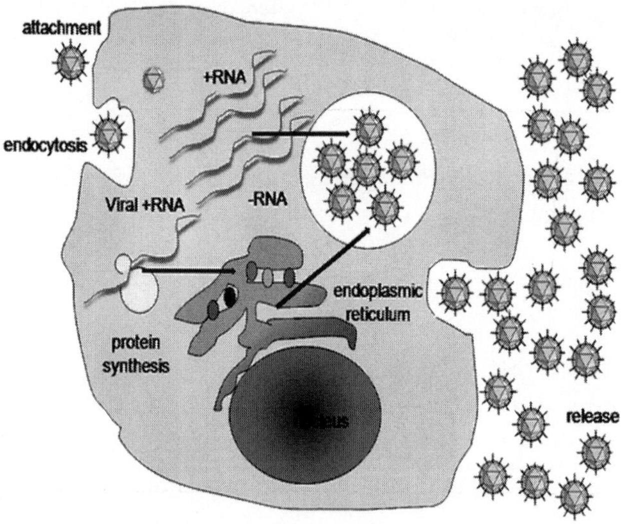

Fig. 11: A typical virus replication cycle

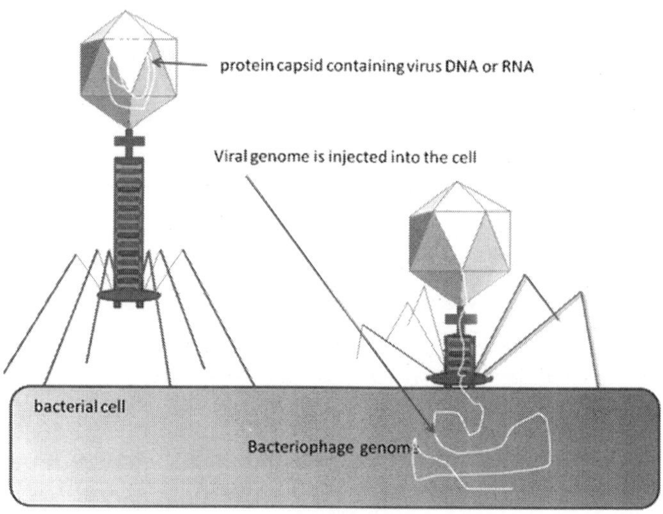

Fig. 12:

Some bacteriophages inject their genomes into bacterial cells

The life cycle of viruses differs greatly between species but there are six basic stages in the life cycle of viruses:

§ Attachment is a specific binding between viral capsid proteins and specific receptors on the host cellular surface. This specificity determines the host range of a virus. For example, HIV infects a limited range of human leucocytes. This is because its surface protein, gp120, specifically interacts with theCD4 molecule – a chemokine receptor – which is most commonly found on the surface of CD4+ T-Cells. This mechanism has evolved to favour those viruses that infect only cells in which they are capable of replication. Attachment to the receptor can induce the viral envelope protein to undergo changes that results in the fusion of viral and cellular membranes, or changes of non-enveloped virus surface proteins that allow the virus to enter.

§ Penetration follows attachment: Virions enter the host cell through receptor-mediated endocytosis or membrane fusion. This is often calledviral entry. The infection of plant and, it is presumed, fungal cells is different from that of animal cells. Plants have a rigid cell wall made ofcellulose, and fungi one of chitin, so most viruses can get inside these cells only after trauma to the cell wall. However, nearly all plant viruses (such as tobacco mosaic virus) can also move directly from cell to cell, in the form of single-stranded nucleoprotein complexes, through pores called plasmodesmata. This process requires movement proteins, which are virus-encoded proteins probably originally derived from plant proteins, which interact with the plasmodesmatal transport machinery Bacteria, like plants, have strong cell walls that a virus must breach to infect the cell. However, given that bacterial cell walls are much less thick than plant cell walls due to their much smaller size, some viruses have evolved mechanisms that inject their genome into the bacterial cell across the cell wall, while the viral capsid remains outside.

§ Uncoating is a process in which the viral capsid is removed: This may be by degradation by viral enzymes or host enzymes or by simple dissociation; the end-result is the releasing of the viral genomic nucleic acid.

§ Replication of viruses primarily involves multiplication of the genome. Replication involves synthesis of viral messenger RNA (mRNA) from "early" genes (with exceptions for positive sense RNA viruses), viral protein synthesis, possible assembly of viral proteins, then viral genome replication mediated by early or regulatory protein expression. This may be followed, for complex viruses with larger genomes, by one or more further rounds of mRNA synthesis: "late" gene expression is, in general, of structural or virion proteins.

§ Following the structure-mediated self-assembly of the virus particles, some modification of the proteins often occurs. In viruses such as HIV, this modification (sometimes called maturation) occurs after the virus has been released from the host cell.

§ Viruses can be released from the host cell by lysis, a process that kills the cell by bursting its membrane and cell wall if present: This is a feature of many bacterial and some animal viruses. Some viruses undergo a lysogenic cycle where the viral genome is incorporated bygenetic recombination into a specific place in the host's chromosome. The viral genome is then known as a "provirus" or, in the case of bacteriophages a "prophage". Whenever the host divides, the viral genome is also replicated. The viral genome is mostly silent within the host; however, at some point, the provirus or prophage may give rise to active virus, which may lyse the host cells. Enveloped viruses (e.g., HIV) typically are released from the host cell by budding. During this process the virus acquires its envelope, which is a modified piece of the host's plasma or other, internal membrane.

The genetic material within virus particles, and the method by which the material is replicated, varies considerably between different types of viruses.

DNA Viruses

The genome replication of most DNA viruses takes place in the cell's nucleus. If the cell has the appropriate receptor on its surface, these viruses enter the cell sometimes by direct fusion with the cell membrane (e.g. herpesviruses) or – more usually – by receptor-mediated endocytosis. Most DNA viruses are entirely dependent on the host cell's DNA and RNA synthesising machinery, and RNA processing machinery; however, viruses with larger genomes may encode much of this machinery themselves. In eukaryotes the viral genome must cross the cell's nuclear membrane to access this machinery, while in bacteria it need only enter the cell.

RNA Viruses

Replication usually takes place in the cytoplasm. RNA viruses can be placed into four different groups depending on their modes of replication. The polarity (whether or not it can be used directly by ribosomes to make proteins) of single-stranded RNA viruses largely determines the replicative mechanism; the other major criterion is whether the genetic material is single-stranded or double-stranded. All RNA viruses use their own RNA replicase enzymes to create copies of their genomes.

Reverse Transcribing Viruses

These have ssRNA (Retroviridae, Metaviridae, Pseudoviridae) or dsDNA (Caulimoviridae, and Hepadnaviridae) in their particles. Reverse transcribing viruses with RNA genomes (retroviruses), use a DNA intermediate to replicate, whereas those with DNA genomes (pararetroviruses) use an RNA intermediate during genome replication. Both types use a reverse transcriptase, or RNA-dependent DNA polymerase enzyme, to carry out the nucleic acid conversion. Retroviruses integrate the DNA produced by reverse transcription into the host genome as a provirus as a part of the replication process; pararetroviruses do not, although integrated genome copies of especially plant pararetroviruses can give rise to infectious virus. They are susceptible to antiviral drugs that inhibit the reverse transcriptase enzyme, e.g. zidovudine and lamivudine. An example of the first type is HIV, which is a retrovirus. Examples of the second type are the Hepadnaviridae, which includes Hepatitis B virus.

Effects on the Host Cell

The range of structural and biochemical effects that viruses have on the host cell is extensive. These are called cytopathic effects. Most virus infections eventually result in the death of the host cell. The causes of death include cell lysis, alterations to the cell's surface membrane and apoptosis. Often cell death is caused by cessation of its normal activities because of suppression by virus-specific proteins, not all of which are components of the virus particle.

Some viruses cause no apparent changes to the infected cell. Cells in which the virus is latent and inactive show few signs of infection and often function normally. This causes persistent infections and the virus is often dormant for many months or years. This is often the case with herpes viruses. Some viruses, such as Epstein-Barr virus, can cause cells to proliferate without causing malignancy, while others, such as papillomaviruses, are established causes of cancer.

Host Range

Viruses are by far the most abundant parasites on earth, and they have been found to infect all types of cellular life including animals, plants, bacteria and fungi. However, different types of viruses can infect only a limited range of hosts and many are species-specific. Some, such as smallpox virus for example, can infect only one species – in this case humans, and are said to have a narrowhost range. Other viruses, such as rabies virus, can infect different species of mammals and are said to have a broad range. The viruses that infect plants are harmless to animals, and most viruses that infect other animals are harmless to humans. The host range of some bacteriophages is limited to a single strain of bacteria and they can be used to trace the source of outbreaks of infections by a method called phage typing.

Classification

Classification seeks to describe the diversity of viruses by naming and grouping them on the basis of similarities. In 1962, André Lwoff, Robert Horne, and Paul Tournier were the first to develop a means of virus classification, based on the Linnaean hierarchical system. This system bases classification on phylum, class, order, family, genus, and species. Viruses were grouped according to their shared properties (not those of their hosts) and the type of nucleic acid forming their genomes. Later the International Committee on Taxonomy of Viruses was formed. However, viruses are not classified on the basis of phylum or class, as their small genome size and high rate of mutation makes it difficult to determine their ancestry beyond Order. As such, the Baltimore Classification is used to supplement the more traditional hierarchy.

Chapter 3

ICTV Classification

The International Committee on Taxonomy of Viruses (ICTV) developed the current classification system and wrote guidelines that put a greater weight on certain virus properties to maintain family uniformity.

A unified taxonomy (a universal system for classifying viruses) has been established. The 7th lCTV Report formalised for the first time the concept of the virus species as the lowest taxon (group) in a branching hierarchy of viral taxa.

However, at present only a small part of the total diversity of viruses has been studied, with analyses of samples from humans finding that about 20% of the virus sequences recovered have not been seen before, and samples from the environment, such as from seawater and ocean sediments, finding that the large majority of sequences are completely novel.

The general taxonomic structure is as follows:

Order (-virales)

Family (-viridae)

Subfamily (-virinae)

Genus (-virus)

Species (-virus)

In the current (2010) ICTV taxonomy, six orders have been established, the Caudovirales, Herpesvirales, Mononegavirales, Nidovirales, Picornavirales and Tymovirales. A seventh order Ligamenvirales has also been proposed. The committee does not formally distinguish between subspecies, strains, and isolates. In total there are 6 orders, 87 families, 19 subfamilies, 348 genera, 2,885 species and about 3,000 types yet unclassified.

Baltimore Classification

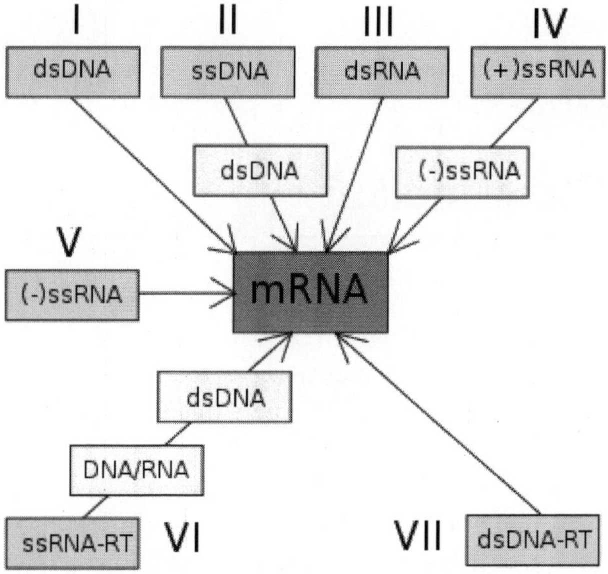

Fig. 1: The Baltimore Classification of viruses is based on the method of viral mRNAsynthesis.

The Nobel Prize-winning biologist David Baltimore devised the Baltimore classification system. The ICTV classification system is used in conjunction with the Baltimore classification system in modern virus classification.

The Baltimore classification of viruses is based on the mechanism of mRNA production. Viruses must generate mRNAs from their genomes to produce proteins and replicate themselves, but different mechanisms are used to achieve this in each virus family. Viral genomes may be single-stranded (ss) or double-stranded (ds), RNA or DNA, and may or may not use reverse transcriptase (RT). Additionally, ssRNA viruses may be either sense (+) or antisense (“). This classification places viruses into seven groups:

- §I: dsDNA viruses (e.g. Adenoviruses, Herpesviruses, Poxviruses)
- §II: ssDNA viruses (+)sense DNA (e.g. Parvoviruses)
- §III: dsRNA viruses (e.g. Reoviruses)
- §IV: (+)ssRNA viruses (+)sense RNA (e.g. Picornaviruses, Togaviruses)
- §V: (“)ssRNA viruses (“)sense RNA (e.g. Orthomyxoviruses, Rhabdoviruses)

§VI: ssRNA-RT viruses (+)sense RNA with DNA intermediate in life-cycle (e.g. Retroviruses)

§VII: dsDNA-RT viruses (e.g. Hepadnaviruses)

As an example of viral classification, the chicken pox virus, varicella zoster (VZV), belongs to the order Herpesvirales, family Herpesviridae, subfamilyAlphaherpesvirinae, and genus Varicellovirus. VZV is in Group I of the Baltimore Classification because it is a dsDNA virus that does not use reverse transcriptase.

Role in Human Disease

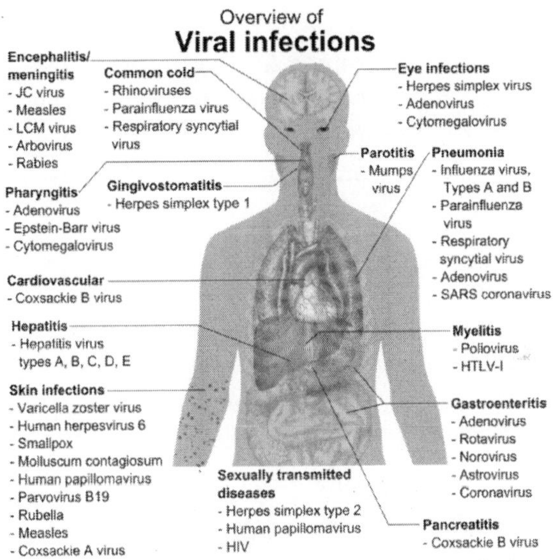

Fig. 2: Overview of the main types of viral infection and the most notable species involved

Examples of common human diseases caused by viruses include the common cold, influenza, chickenpox and cold sores. Many serious diseases such asebola, AIDS, avian influenza and SARS are caused by viruses. The relative ability of viruses to cause disease is described in terms of virulence.

Other diseases are under investigation as to whether they too have a virus as the causative agent, such as the possible connection between human herpes virus six(HHV6) and neurological diseases such as multiple sclerosis and chronic fatigue syndrome. There is controversy over whether the borna virus, previously thought to cause neurological diseases in horses, could be responsible for psychiatric illnesses in humans.

Viruses have different mechanisms by which they produce disease in an organism, which largely depends on the viral species. Mechanisms at the cellular level primarily include cell lysis, the breaking open and subsequent death of the cell. In multicellular organisms, if enough cells die, the whole organism will start to suffer the effects. Although viruses cause disruption of healthy homeostasis, resulting in disease, they may exist relatively harmlessly within an organism. An example would include the ability of the herpes simplex virus, which causes cold sores, to remain in a dormant state within the human body. This is called latency and is a characteristic of the herpes viruses including Epstein-Barr virus, which causes glandular fever, and varicella zoster virus, which causes chickenpox and shingles. Most people have been infected with at least one of these types of herpes virus. However, these latent viruses might sometimes be beneficial, as the presence of the virus can increase immunity against bacterial pathogens, such as Yersinia pestis.

Some viruses can cause life-long or chronic infections, where the viruses continue to replicate in the body despite the host's defence mechanisms. This is common in hepatitis B virus and hepatitis C virus infections. People chronically infected are known as carriers, as they serve as reservoirs of infectious virus. In populations with a high proportion of carriers, the disease is said to be endemic.

Epidemiology

Viral epidemiology is the branch of medical science that deals with the transmission and control of virus infections in humans. Transmission of viruses can be vertical, that is from mother to child, or horizontal, which means from person to person. Examples of vertical transmission include hepatitis B virus and HIV where the baby is born already infected with the virus. Another, more rare, example is the varicella zoster virus, which, although causing relatively mild infections in humans, can be fatal to the foetus and new-born baby.

Horizontal transmission is the most common mechanism of spread of viruses in populations. Transmission can occur when: body fluids are exchanged during sexual activity, e.g., HIV; blood is exchanged by contaminated transfusion or needle sharing, e.g., hepatitis C; a child is born to an infected mother, e.g., hepatitis B; exchange of saliva by mouth, e.g., Epstein-Barr virus; contaminated food or water is ingested, e.g., norovirus; aerosols containing virions are inhaled, e.g., influenza virus; and insect vectors such as mosquitoes penetrate the skin of a host, e.g., dengue. The rate or speed of transmission of viral infections depends on factors that include population density, the number of susceptible individuals, (i.e., those not immune), the quality of healthcare and the weather.

Epidemiology is used to break the chain of infection in populations during outbreaks of viral diseases. Control measures are used that are based on knowledge of how the virus is transmitted. It is important to find the source, or sources, of the outbreak and to identify the virus. Once the virus has been identified, the chain of transmission can sometimes be broken by vaccines. When vaccines are not available sanitation and disinfection can be effective. Often infected people are isolated from the rest of the community and those that have been exposed to the virus placed in quarantine. To control the outbreak of foot and mouth disease in cattle in Britain in 2001, thousands of cattle were slaughtered. Most viral infections of humans and other animals have incubation periods during which the infection causes no signs or symptoms. Incubation periods for viral diseases range from a few days to weeks but are known for most infections. Somewhat overlapping, but mainly following the incubation period, there is a period of communicability; a time when an infected individual or animal is contagious and can infect another person or animal. This too is known for many viral infections and knowledge the length of both periods is important in the control of outbreaks. When outbreaks cause an unusually high proportion of cases in a population, community or region they are called epidemics. If outbreaks spread worldwide they are called pandemics.

Epidemics and Pandemics

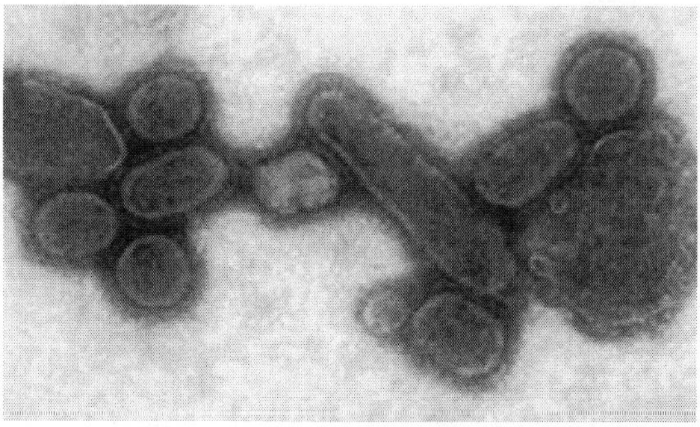

Fig. 3: Transmission electron microscope image of a recreated 1918 influenza virus

Native American populations were devastated by contagious diseases, in particular, smallpox, brought to the Americas by European colonists. It is unclear how many Native Americans were killed by foreign diseases after the arrival of Columbus in the Americas, but the numbers have been estimated to be close to 70% of the indigenous population. The damage done by this disease significantly aided European attempts to displace and conquer the native population.

A pandemic is a worldwide epidemic. The 1918 flu pandemic, commonly referred to as the Spanish flu, was a category 5 influenza pandemic caused by an unusually severe and deadly influenza A virus. The victims were often healthy young adults, in contrast to most influenza outbreaks, which predominantly affect juvenile, elderly, or otherwise-weakened patients.

The Spanish flu pandemic lasted from 1918 to 1919. Older estimates say it killed 40–50 million people, while more recent research suggests that it may have killed as many as 100 million people, or 5% of the world's population in 1918. Most researchers believe that HIV originated in sub-Saharan Africaduring the 20th century; it is now a pandemic, with an estimated 38.6 million people now living with the disease worldwide.

The Joint United Nations Programme on HIV/AIDS (UNAIDS) and the World Health Organization (WHO) estimate that AIDS has killed more than 25 million people since it was first recognised on June 5, 1981, making it one of the most destructive epidemics in recorded history. In 2007 there were 2.7 million new HIV infections and 2 million HIV-related deaths.

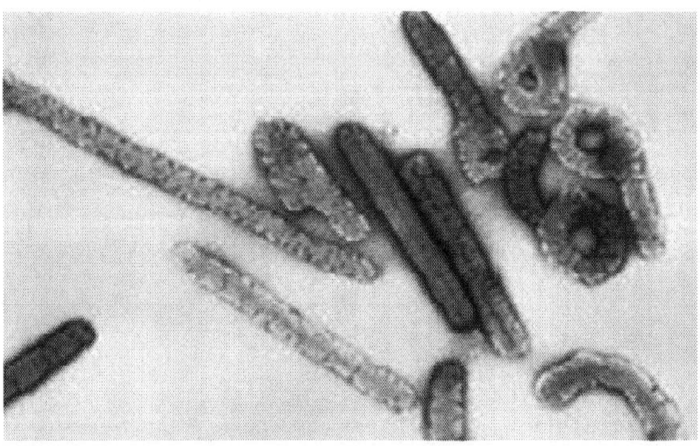

Fig. 4: Marburg virus

Several highly lethal viral pathogens are members of the Filoviridae. Filoviruses are filament-like viruses that cause viral hemorrhagic fever, and include theebola and marburg viruses. The Marburg virus attracted widespread press attention in April 2005 for an outbreak in Angola. Beginning in October 2004 and continuing into 2005, the outbreak was the world's worst epidemic of any kind of viral hemorrhagic fever.

Cancer

Viruses are an established cause of cancer in humans and other species. Viral cancers occur only in a minority of infected persons (or animals). Cancer viruses come from a range of virus families, including both RNA and DNA viruses, and so there is no single type of "oncovirus" (an obsolete term originally used for acutely transforming retroviruses). The development of cancer is determined by a variety of factors such as host immunity and mutations in the host.

Viruses accepted to cause human cancers include some genotypes of human papillomavirus, hepatitis B virus, hepatitis C virus, Epstein-Barr virus,Kaposi's sarcoma-associated herpesvirus and human T-lymphotropic virus. The most recently discovered human cancer virus is a polyomavirus (Merkel cell polyomavirus) that causes most cases of a rare form of skin cancer called Merkel cell carcinoma.

Hepatitis viruses can develop into a chronic viral infection that leads to liver cancer.Infection by human T-lymphotropic virus can lead to tropical spastic paraparesis and adult T-cell leukemia. Human papillomaviruses are an established cause of cancers of cervix, skin, anus, andpenis.

Within the Herpesviridae, Kaposi's sarcoma-associated herpesvirus causes Kaposi's sarcoma and body cavity lymphoma, and Epstein–Barr virus causes Burkitt's lymphoma, Hodgkin's lymphoma, B lymphoproliferative disorder, and nasopharyngeal carcinoma. Merkel cell polyomavirus closely related to SV40 and mouse polyomaviruses that have been used as animal models for cancer viruses for over 50 years.

Host Defence Mechanisms

The body's first line of defence against viruses is the innate immune system. This comprises cells and other mechanisms that defend the host from infection in a non-specific manner. This means that the cells of the innate system recognise, and respond to, pathogens in a generic way, but, unlike the adaptive immune system, it does not confer long-lasting or protective immunity to the host.

RNA interference is an important innate defence against viruses. Many viruses have a replication strategy that involves double-stranded RNA (dsRNA). When such a virus infects a cell, it releases its RNA molecule or molecules, which immediately bind to a protein complex called dicer that cuts the RNA into smaller pieces. A biochemical pathway called the RISC complex is activated, which degrades the viral mRNA and the cell survives the infection. Rotaviruses avoid this mechanism by not uncoating fully inside the cell and by releasing newly produced mRNA through pores in the particle's inner capsid. The genomic dsRNA remains protected inside the core of the virion.

When the adaptive immune system of a vertebrate encounters a virus, it produces specific antibodies that bind to the virus and render it non-infectious. This is called humoral immunity. Two types of antibodies are important. The first, called IgM, is highly effective at neutralizing viruses but is produced by the cells of the immune system only for a few weeks. The second, called IgG, is produced indefinitely. The presence of IgM in the blood of the host is used to test for acute infection, whereas IgG indicates an infection sometime in the past. IgG antibody is measured when tests forimmunity are carried out.

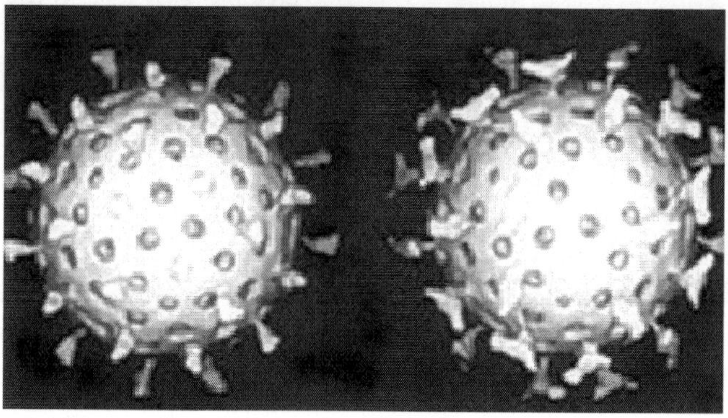

Fig. 5: Two rotaviruses: the one on the right is coated with antibodies that stop its attaching to cells and infecting them

A second defence of vertebrates against viruses is called cell-mediated immunity and involves immune cells known as T cells. The body's cells constantly display short fragments of their proteins on the cell's surface, and, if a T cell recognises a suspicious viral fragment there, the host cell is destroyed by killer Tcells and the virus-specific T-cells proliferate. Cells such as the macrophage are specialists at this antigen presentation. The production of interferon is an important host defence mechanism. This is a hormone produced by the body when viruses are present. Its role in immunity is complex; it eventually stops the viruses from reproducing by killing the infected cell and its close neighbours.

Not all virus infections produce a protective immune response in this way. HIV evades the immune system by constantly changing the amino acid sequence of the proteins on the surface of the virion. These persistent viruses evade immune control by sequestration, blockade of antigen presentation, cytokineresistance, evasion of natural killer cell activities, escape from apoptosis, and antigenic shift. Other viruses, called neurotropic viruses, are disseminated by neural spread where the immune system may be unable to reach them.

Prevention and Treatment

Because viruses use vital metabolic pathways within host cells to replicate, they are difficult to eliminate without using drugs that cause toxic effects to host cells in general. The most effective medical approaches to viral diseases are vaccinations to provide immunity to infection, and antiviral drugs that selectively interfere with viral replication.

Vaccines

Vaccination is a cheap and effective way of preventing infections by viruses. Vaccines were used to prevent viral infections long before the discovery of the actual viruses. Their use has resulted in a dramatic decline in morbidity (illness) and mortality (death) associated with viral infections such as polio, measles, mumps and rubella. Smallpox infections have been eradicated. Vaccines are available to prevent over thirteen viral infections of humans, and more are used to prevent viral infections of animals. Vaccines can consist of live-attenuated or killed viruses, or viral proteins (antigens). Live vaccines contain weakened forms of the virus, which do not cause the disease but, nonetheless, confer immunity.

Such viruses are called attenuated. Live vaccines can be dangerous when given to people with a weak immunity, (who are described as immunocompromised), because in these people, the weakened virus can cause the original disease. Biotechnology and genetic engineering techniques are used to produce subunit vaccines. These vaccines use only the capsid proteins of the virus. Hepatitis B vaccine is an example of this type of vaccine.Subunit vaccines are safe for immuno-compromised patients because they cannot cause the disease. The yellow fever virus vaccine, a live-attenuated strain called 17D, is probably the safest and most effective vaccine ever generated.

Antiviral Drugs

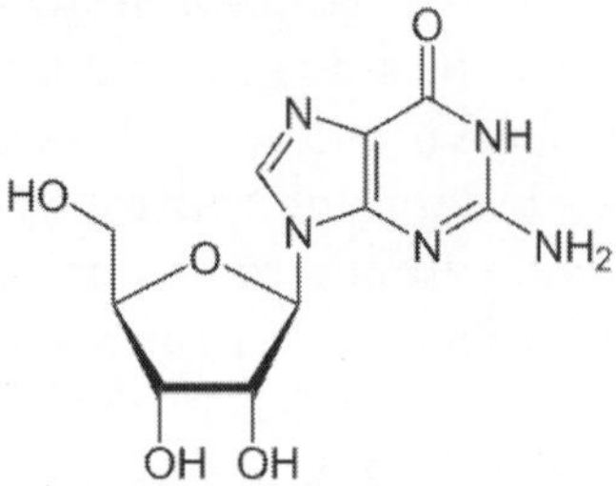

Fig. 9: Guanosine

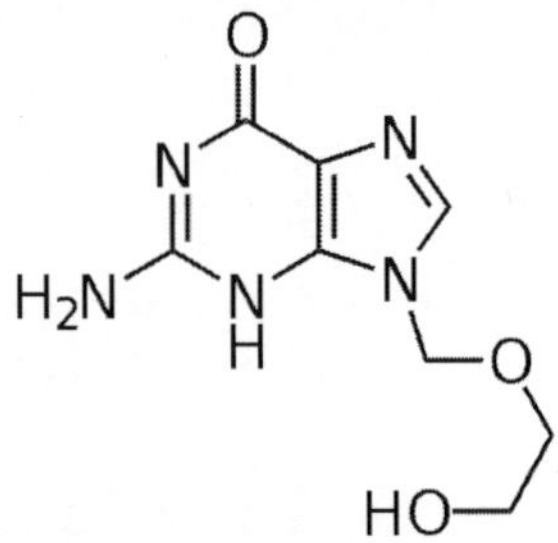

Fig. 10: The guanosine analogue Aciclovir

Antiviral drugs are often nucleoside analogues, (fake DNA building-blocks), which viruses mistakenly incorporate into their genomes during replication. The life-cycle of the virus is then halted because the newly synthesised DNA is inactive. This is because these analogues lack the hydroxyl groups, which, along withphosphorus atoms, link together to form the strong "backbone" of the DNA molecule. This is called DNA chain termination.

Examples of nucleoside analogues are aciclovir for Herpes simplex virus infections and lamivudine for HIV and Hepatitis B virus infections. Aciclovir is one of the oldest and most frequently prescribed antiviral drugs. Other antiviral drugs in use target different stages of the viral life cycle. HIV is dependent on a proteolytic enzyme called the HIV-1 protease for it to become fully infectious. There is a large class of drugs called protease inhibitors that inactivate this enzyme.

Hepatitis C is caused by an RNA virus. In 80% of people infected, the disease is chronic, and without treatment, they are infected for the remainder of their lives. However, there is now an effective treatment that uses the nucleoside analogue drug ribavirin combined with interferon. The treatment of chroniccarriers of the hepatitis B virus by using a similar strategy using lamivudine has been developed.

Infection in Other Species

Viruses infect all cellular life and, although viruses occur universally, each cellular species has its own specific range that often infect only that species. Some viruses, called satellites, can only replicate within cells that have already been infected by another virus. Viruses are important pathogens of livestock. Diseases such as Foot and Mouth Disease and bluetongue are caused by viruses. Companion animals such as cats, dogs, and horses, if not vaccinated, are susceptible to serious viral infections. Canine parvovirus is caused by a small DNA virus and infections are often fatal in pups. Like all invertebrates, the honey bee is susceptible to many viral infections. Fortunately, most viruses co-exist harmlessly in their host and cause no signs or symptoms of disease.

Plants

Fig. 11: Peppers infected by mild mottle virus

There are many types of plant virus, but often they cause only a loss of yield, and it is not economically viable to try to control them. Plant viruses are often spread from plant to plant by organisms, known as vectors. These are normally insects, but some fungi, nematode worms, and single-celled organisms have been shown to be vectors.

When control of plant virus infections is considered economical, for perennial fruits, for example, efforts are concentrated on killing the vectors and removing alternate hosts such as weeds. Plant viruses are harmless to humans and other animals because they can reproduce only in living plant cells.

Plants have elaborate and effective defence mechanisms against viruses. One of the most effective is the presence of so-called resistance (R) genes. Each R gene confers resistance to a particular virus by triggering localised areas of cell death around the infected cell, which can often be seen with the unaided eye as large spots. This stops the infection from spreading. RNA interference is also an effective defence in plants. When they are infected, plants often produce natural disinfectants that kill viruses, such as salicylic acid, nitric oxide, and reactive oxygen molecules.

Plant virus particles or virus-like particles (VLPs) have applications in both biotechnology and nanotechnology. The capsids of most plant viruses are simple and robust structures and can be produced in large quantities either by the infection of plants or by expression in a variety of heterologous systems. Plant virus particles can be modified genetically and chemically to encapsulate foreign material and can be incorporated into supramolecular structures for use in biotechnology.

Bacteria

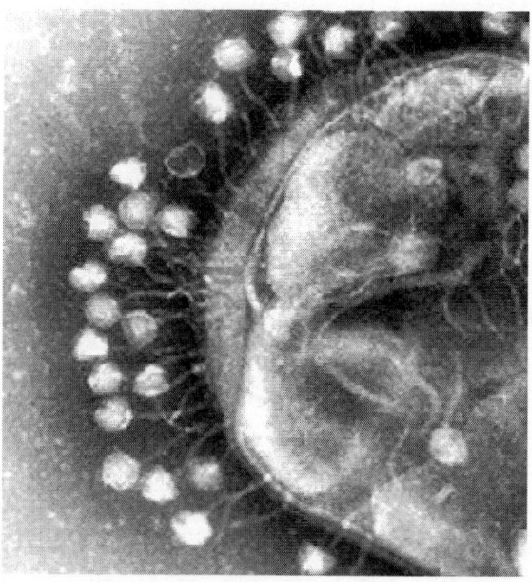

Fig. 12: Transmission electron micrograph of multiple bacteriophages attached to a bacterial cell wall

Bacteriophages are a common and diverse group of viruses and are the most abundant form of biological entity in aquatic environments – there are up to ten times more of these viruses in the oceans than there are bacteria, reaching levels of 250,000,000 bacteriophages per millilitre of seawater. These viruses infect specific bacteria by binding to surface receptor molecules and then entering the cell. Within a short amount of time, in some cases just minutes, bacterial polymerase starts translating viral mRNA into protein. These proteins go on to become either new virions within the cell, helper proteins, which help assembly of new virions, or proteins involved in cell lysis. Viral enzymes aid in the breakdown of the cell membrane, and, in the case of the T4 phage, in just over twenty minutes after injection over three hundred phages could be released.

The major way bacteria defend themselves from bacteriophages is by producing enzymes that destroy foreign DNA. These enzymes, called restriction endonucleases, cut up the viral DNA that bacteriophages inject into bacterial cells. Bacteria also contain a system that uses CRISPR sequences to retain fragments of the genomes of viruses that the bacteria have come into contact with in the past, which allows them to block the virus's replication through a form of RNA interference. This genetic system provides bacteria with acquired immunity to infection.

Archaea

Some viruses replicate within archaea: these are double-stranded DNA viruses with unusual and sometimes unique shapes. These viruses have been studied in most detail in the thermophilic archaea, particularly the orders Sulfolobales and Thermoproteales. Defences against these viruses may involveRNA interference from repetitive DNA sequences within archaean genomes that are related to the genes of the viruses.

Role in Aquatic Ecosystems

Viruses are the most abundant biological entity in aquatic environments: a teaspoon of seawater contains about one million of them. They are essential to the regulation of saltwater and freshwater ecosystems. Most of these viruses are bacteriophages, which are harmless to plants and animals. They infect and destroy the bacteria in aquatic microbial communities, comprising the most important mechanism of recycling carbon in the marine environment. The organic molecules released from the bacterial cells by the viruses stimulates fresh bacterial and algal growth. Microorganisms constitute more than 90% of the biomass in the sea. It is estimated that viruses kill approximately 20% of this biomass each day and that there are 15 times as many viruses in the oceans as there are bacteria and archaea. Viruses are the main agents responsible for the rapid destruction of harmful algal blooms, which often kill other marine life. The number of viruses in the oceans decreases further offshore and deeper into the water, where there are fewer host organisms.

The effects of marine viruses are far-reaching; by increasing the amount of photosynthesis in the oceans, viruses are indirectly responsible for reducing the amount of carbon dioxide in the atmosphere by approximately 3 gigatonnes of carbon per year. Like any organism, marine mammals are susceptible to viral infections. In 1988 and 2002, thousands of harbour seals were killed in Europe by phocine distemper virus. Many other viruses, including caliciviruses, herpesviruses, adenoviruses and parvoviruses, circulate in marine mammal populations.

Role in Evolution

Viruses are an important natural means of transferring genes between different species, which increases genetic diversity and drives evolution. It is thought that viruses played a central role in the early evolution, before the diversification of bacteria, archaea and eukaryotes and at the time of the last universal common ancestor of life on Earth. Viruses are still one of the largest reservoirs of unexplored genetic diversity on Earth.

Applications

Life Sciences and Medicine

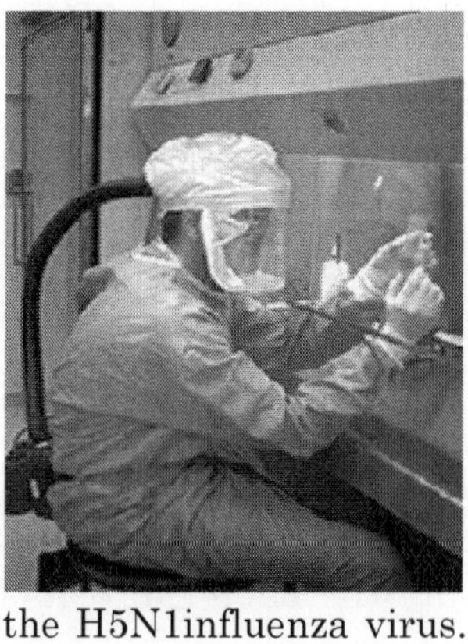

Fig. 13: Scientist studying the H5N1influenza virus.

Viruses are important to the study of molecular and cellular biology as they provide simple systems that can be used to manipulate and investigate the functions of cells. The study and use of viruses have provided valuable information about aspects of cell biology. For example, viruses have been useful in the study ofgenetics and helped our understanding of the basic mechanisms of molecular genetics, such as DNA replication, transcription, RNA processing, translation, proteintransport, and immunology.

Geneticists often use viruses as vectors to introduce genes into cells that they are studying. This is useful for making the cell produce a foreign substance, or to study the effect of introducing a new gene into the genome.

In similar fashion, virotherapy uses viruses as vectors to treat various diseases, as they can specifically target cells and DNA. It shows promising use in the treatment of cancer and in gene therapy. Eastern European scientists have used phage therapy as an alternative to antibiotics for some time, and interest in this approach is increasing, because of the high level of antibiotic resistance now found in some pathogenic bacteria.

Expression of heterologous proteins by viruses is the basis of several manufacturing processes that are currently being used for the production of various proteins such as vaccine antigens and antibodies. Industrial processes have been recently developed using viral vectors and a number of pharmaceutical proteins are currently in pre-clinical and clinical trials.

Materials Science and Nanotechnology

Current trends in nanotechnology promise to make much more versatile use of viruses. From the viewpoint of a materials scientist, viruses can be regarded as organic nanoparticles. Their surface carries specific tools designed to cross the barriers of their host cells. The size and shape of viruses, and the number and nature of the functional groups on their surface, is precisely defined. As such, viruses are commonly used in materials science as scaffolds for covalently linked surface modifications. A particular quality of viruses is that they can be tailored by directed evolution. The powerful techniques developed by life sciences are becoming the basis of engineering approaches towards nanomaterials, opening a wide range of applications far beyond biology and medicine.

Because of their size, shape, and well-defined chemical structures, viruses have been used as templates for organizing materials on the nanoscale.

Recent examples include work at the Naval Research Laboratory in Washington, DC, using Cowpea Mosaic Virus (CPMV) particles to amplify signals in DNA microarray based sensors. In this application, the virus particles separate thefluorescent dyes used for signalling to prevent the formation of non-fluorescent dimers that act as quenchers. Another example is the use of CPMV as a nanoscale breadboard for molecular electronics.

Synthetic Viruses

Many viruses can be synthesized de novo ("from scratch") and the first synthetic virus was created in 2002. Although somewhat of a misconception, it is not the actual virus that is synthesized, but rather its DNA genome (in case of a DNA virus), or a cDNA copy of its genome (in case of RNA viruses).

For many virus families the naked synthetic DNA or RNA (once enzymatically converted back from the synthetic cDNA) is infectious when introduced into a cell. That is, they contain all the necessary information to produce new viruses. This technology is now being used to investigate novel vaccine strategies.

The ability to synthesize viruses has far-reaching consequences, since viruses can no longer be regarded as extinct, as long as the information of their genome sequence is known and permissive cells are available. Currently, the full-length genome sequences of 2408 different viruses (including smallpox) are publicly available at an online database, maintained by theNational Institute of Health.

Weapons

The ability of viruses to cause devastating epidemics in human societies has led to the concern that viruses could be weaponised for biological warfare. Further concern was raised by the successful recreation of the infamous 1918 influenza virus in a laboratory.

The smallpox virus devastated numerous societies throughout history before its eradication. There are officially only two centers in the world that keep stocks of smallpox virus – the Russian Vector laboratory, and the United States Centers for Disease Control.

But fears that it may be used as a weapon are not totally unfounded; the vaccine for smallpox has sometimes severe side-effects – during the last years before the eradication of smallpox disease more people became seriously ill as a result of vaccination than did people from smallpox – and smallpox vaccination is no longer universally practiced.

Thus, much of the modern human population has almost no established resistance to smallpox.

Gene

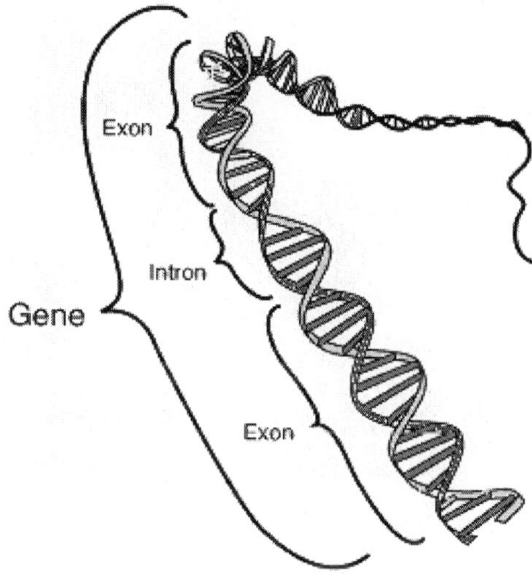

Fig. 14:

A gene is a molecular unit of heredity of a living organism. It is a name given to some stretches of DNA and RNA that code for a type of protein or for an RNA chain that has a function in the organism. Living beings depend on genes, as they specify all proteins and functional RNA chains.

Genes hold the information to build and maintain an organism's cells and pass genetic traits to offspring, although some organelles (e.g. mitochondria) are self-replicating and are not coded for by the organism's DNA.

All organisms have many genes corresponding to various different biological traits, some of which are immediately visible, such as eye color or number of limbs, and some of which are not, such as blood type or increased risk for specific diseases, or the thousands of basic biochemical processes that comprise life.

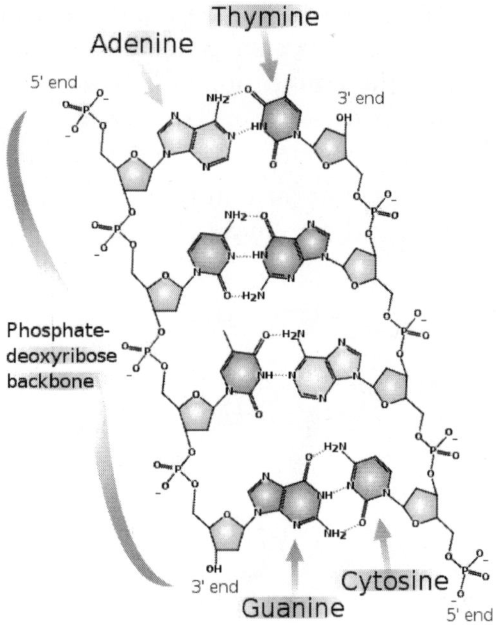

Fig. 15: The chemical structure of a four-base fragment of a DNAdouble helix.

A modern working definition of a gene is "a locatable region of genomic sequence, corresponding to a unit of inheritance, which is associated with regulatory regions, transcribed regions, and or other functional sequence regions ". Colloquial usage of the term gene (e.g. "good genes", "hair color gene") may actually refer to an allele: a gene is the basic instruction — a sequence of nucleic acids (DNA or, in the case of certain virusesRNA), while an allele is one variant of that gene. Referring to having a gene for a trait is no longer the scientifically accepted usage. In most cases, all people would have a gene for the trait in question, but certain people will have a specific allele of that gene, which results in the trait variant. Further, genes code for proteins, which might result in identifiable traits, but it is the gene, not the trait, which is inherited.

Physical Definitions

RNA Genes and Genomes

When proteins are manufactured, the gene is first copied into RNA as an intermediate product. In other cases, the RNA molecules are the actual functional products. For example, RNAs known asribozymes are capable of enzymatic function, and microRNA has a regulatory role. The DNA sequences from which such RNAs are transcribed are known as RNA genes.

Some viruses store their entire genomes in the form of RNA, and contain no DNA at all. Because they use RNA to store genes, their cellular hosts may synthesize their proteins as soon as they areinfected and without the delay in waiting for transcription. On the other hand, RNA retroviruses, such as HIV, require the reverse transcription of their genome from RNA into DNA before their proteins can be synthesized. In 2006, French researchers came across a puzzling example of RNA-mediated inheritance in mice. Mice with a loss-of-function mutation in the gene Kit have white tails. Offspring of these mutants can have white tails despite having only normal Kit genes. The research team traced this effect back to mutated Kit RNA. While RNA is common as genetic storage material in viruses, in mammals in particular RNA inheritance has been observed very rarely.

Functional Structure of a Gene

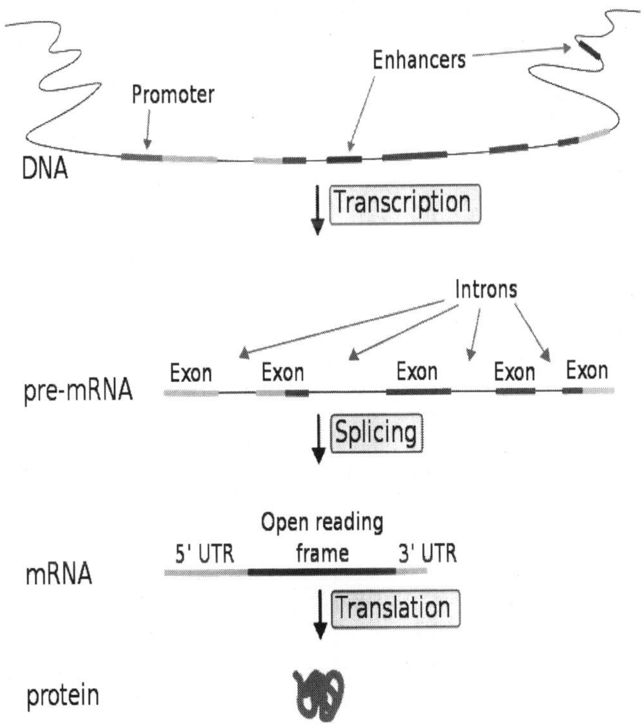

FIg. 16

Diagram of the "typical" eukaryotic protein-coding gene. Promoters andenhancers determine what portions of the DNA will be transcribed into theprecursor mRNA (pre-mRNA). The pre-mRNA is then spliced into messenger RNA(mRNA) which is later translated into protein.

The vast majority of living organisms encode their genes in long strands of DNA. DNA (deoxyribonucleic acid) consists of a chain made from four types of nucleotide subunits, each composed of: a five-carbon sugar (2'-deoxyribose), a phosphategroup, and one of the four bases adenine, cytosine, guanine, and thymine. The most common form of DNA in a cell is in adouble helix structure, in which two individual DNA strands twist around each other in a right-handed spiral. In this structure, the base pairing rules specify that guanine pairs with cytosine and adenine pairs with thymine.

The base pairing between guanine and cytosine forms three hydrogen bonds, whereas the base pairing between adenine and thymine forms two hydrogen bonds. The two strands in a double helix must therefore be complementary, that is, their bases must align such that the adenines of one strand are paired with the thymines of the other strand, and so on.

Due to the chemical composition of the pentose residues of the bases, DNA strands have directionality. One end of a DNA polymer contains an exposed hydroxyl group on the deoxyribose; this is known as the 3' end of the molecule. The other end contains an exposed phosphate group; this is the 5' end. The directionality of DNA is vitally important to many cellular processes, since double helices are necessarily directional (a strand running 5'-3' pairs with a complementary strand running 3'-5'), and processes such as DNA replication occur in only one direction. All nucleic acid synthesis in a cell occurs in the 5'-3' direction, because new monomers are added via a dehydration reaction that uses the exposed 3' hydroxyl as a nucleophile.

The expression of genes encoded in DNA begins by transcribing the gene into RNA, a second type of nucleic acid that is very similar to DNA, but whose monomers contain the sugar ribose rather than deoxyribose. RNA also contains the base uracil in place of thymine. RNA molecules are less stable than DNA and are typically single-stranded. Genes that encode proteins are composed of a series of three-nucleotide sequences called codons, which serve as the words in the genetic language. Thegenetic code specifies the correspondence during protein translation between codons and amino acids. The genetic code is nearly the same for all known organisms.

All genes have regulatory regions in addition to regions that explicitly code for a protein or RNA product. A regulatory regionshared by almost all genes is known as the promoter, which provides a position that is recognized by the transcription machinery when a gene is about to be transcribed and expressed. A gene can have more than one promoter, resulting in RNAs that differ in how far they extend in the 5' end. Although promoter regions have a consensus sequence that is the most common sequence at this position, some genes have "strong" promoters that bind the transcription machinery well, and others have "weak" promoters that bind poorly. These weak promoters usually permit a lower rate of transcription than the strong promoters, because the transcription machinery binds to them and initiates transcription less frequently. Other possible regulatory regions include enhancers, which can compensate for a weak promoter. Most regulatory regions are "upstream"—that is, before or toward the 5' end of the transcription initiation site. Eukaryotic promoter regions are much more complex and difficult to identify than prokaryotic promoters.

Many prokaryotic genes are organized into operons, or groups of genes whose products have related functions and which are transcribed as a unit. By contrast, eukaryotic genes are transcribed only one at a time, but may include long stretches of DNA called introns which are transcribed but never translated into protein (they are spliced out before translation). Splicing can also occur in prokaryotic genes, but is less common than in eukaryotes.

Chapter 4

Chromosomes

The total complement of genes in an organism or cell is known as its genome, which may be stored on one or more chromosomes; the region of the chromosome at which a particular gene is located is called its locus. A chromosome consists of a single, very long DNA helix on which thousands of genes are encoded. Prokaryotes—bacteria and archaea—typically store their genomes on a single large, circular chromosome, sometimes supplemented by additional small circles of DNA called plasmids, which usually encode only a few genes and are easily transferable between individuals.

For example, the genes for antibiotic resistance are usually encoded on bacterial plasmids and can be passed between individual cells, even those of different species, via horizontal gene transfer. Although some simple eukaryotes also possess plasmids with small numbers of genes, the majority of eukaryotic genes are stored on multiple linear chromosomes, which are packed within the nucleus in complex with storage proteins called histones. The manner in which DNA is stored on the histone, as well as chemical modifications of the histone itself, are regulatory mechanisms governing whether a particular region of DNA is accessible for gene expression. The ends of eukaryotic chromosomes are capped by long stretches of repetitive sequences called telomeres, which do not code for any gene product but are present to prevent degradation of coding and regulatory regions during DNA replication. The length of the telomeres tends to decrease each time the genome is replicated in preparation for cell division; the loss of telomeres has been proposed as an explanation for cellular senescence, or the loss of the ability to divide, and by extension for the aging process in organisms.

Whereas the chromosomes of prokaryotes are relatively gene-dense, those of eukaryotes often contain so-called "junk DNA", or regions of DNA that serve no obvious function. Simple single-celled eukaryotes have relatively small amounts of such DNA, whereas the genomes of complex multicellular organisms, including humans, contain an absolute majority of DNA without an identified function. However it now appears that, although protein-coding DNA makes up barely 2% of the human genome, about 80% of the bases in the genome may be expressed, so the term "junk DNA" may be a misnomer.

Gene Expression

In all organisms, there are two major steps separating a protein-coding gene from its protein: First, the DNA on which the gene resides must be transcribed from DNA to messenger RNA (mRNA); and, second, it must be translated from mRNA to protein. RNA-coding genes must still go through the first step, but are not translated into protein. The process of producing a biologically functional molecule of either RNA or protein is called gene expression, and the resulting molecule itself is called a gene product.

Genetic Code

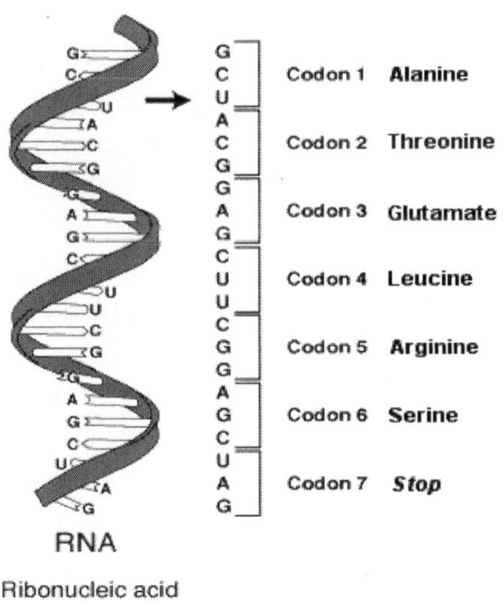

Fig. 1: Schematic diagram of a single-stranded RNA molecule illustrating the position of three-base codons.

The genetic code is the set of rules by which a gene is translated into a functional protein. Each gene consists of a specific sequence of nucleotides encoded in a DNA (or sometimes RNA in some viruses) strand; a correspondence between nucleotides, the basic building blocks of genetic material, and amino acids, the basic building blocks of proteins, must be established for genes to be successfully translated into functional proteins. Sets of three nucleotides, known ascodons, each correspond to a specific amino acid or to a signal; three codons are known as "stop codons" and, instead of specifying a new amino acid, alert the translation machinery that the end of the gene has been reached.

There are 64 possible codons (four possible nucleotides at each of three positions, hence 4possible codons) and only 20 standard amino acids; hence the code is redundant and multiple codons can specify the same amino acid. The correspondence between codons and amino acids is nearly universal among all known living organisms.

Transcription

The process of genetic transcription produces a single-stranded RNA molecule known as messenger RNA, whose nucleotide sequence is complementary to the DNA from which it was transcribed. The DNA strand whose sequence matches that of the RNA is known as the coding strand and the strand from which the RNA was synthesized is the template strand.

Transcription is performed by an enzyme called an RNA polymerase, which reads the template strand in the 3' to 5'direction and synthesizes the RNA from 5' to 3'. To initiate transcription, the polymerase first recognizes and binds a promoter region of the gene. Thus a major mechanism of gene regulation is the blocking or sequestering of the promoter region, either by tight binding by repressor molecules that physically block the polymerase, or by organizing the DNA so that the promoter region is not accessible.

In prokaryotes, transcription occurs in the cytoplasm; for very long transcripts, translation may begin at the 5' end of the RNA while the 3' end is still being transcribed. In eukaryotes, transcription necessarily occurs in the nucleus, where the cell's DNA is sequestered; the RNA molecule produced by the polymerase is known as the primary transcript and must undergo post-transcriptional modifications before being exported to the cytoplasm for translation.

The splicing ofintrons present within the transcribed region is a modification unique to eukaryotes; alternative splicing mechanisms can result in mature transcripts from the same gene having different sequences and thus coding for different proteins. This is a major form of regulation in eukaryotic cells.

Translation

DNA Replication and Inheritance

The growth, development, and reproduction of organisms relies on cell division, or the process by which a single cell divides into two usually identical daughter cells. This requires first making a duplicate copy of every gene in the genome in a process called DNA replication. The copies are made by specialized enzymes known as DNA polymerases, which "read" one strand of the double-helical DNA, known as the template strand, and synthesize a new complementary strand. Because the DNA double helix is held together by base pairing, the sequence of one strand completely specifies the sequence of its complement; hence only one strand needs to be read by the enzyme to produce a faithful copy. The process of DNA replication is semiconservative; that is, the copy of the genome inherited by each daughter cell contains one original and one newly synthesized strand of DNA.

After DNA replication is complete, the cell must physically separate the two copies of the genome and divide into two distinct membrane-bound cells. In prokaryotes - bacteria and archaea - this usually occurs via a relatively simple process called binary fission, in which each circular genome attaches to the cell membrane and is separated into the daughter cells as the membrane invaginates to split the cytoplasm into two membrane-bound portions. Binary fission is extremely fast compared to the rates of cell division in eukaryotes. Eukaryotic cell division is a more complex process known as thecell cycle; DNA replication occurs during a phase of this cycle known as S phase, whereas the process of segregating chromosomes and splitting the cytoplasm occurs during M phase. In many single-celled eukaryotes such as yeast, reproduction by budding is common, which results in asymmetrical portions of cytoplasm in the two daughter cells.

Molecular Inheritance

The duplication and transmission of genetic material from one generation of cells to the next is the basis for molecular inheritance, and the link between the classical and molecular pictures of genes. Organisms inherit the characteristics of their parents because the cells of the offspring contain copies of the genes in their parents' cells. In asexually reproducing organisms, the offspring will be a genetic copy or clone of the parent organism. In sexually reproducing organisms, a specialized form of cell division called meiosis produces cells called gametes or germ cells that are haploid, or contain only one copy of each gene. The gametes produced by females are called eggs or ova, and those produced by males are called sperm. Two gametes fuse to form a fertilized egg, a single cell that once again has a diploid number of genes—each with one copy from the mother and one copy from the father.

During the process of meiotic cell division, an event called genetic recombination or crossing-over can sometimes occur, in which a length of DNA on one chromatid is swapped with a length of DNA on the corresponding sister chromatid. This has no effect if the alleles on the chromatids are the same, but results in reassortment of otherwise linked alleles if they are different. The Mendelian principle of independent assortment asserts that each of a parent's two genes for each trait will sort independently into gametes; which allele an organism inherits for one trait is unrelated to which allele it inherits for another trait.

This is in fact only true for genes that do not reside on the same chromosome, or are located very far from one another on the same chromosome. The closer two genes lie on the same chromosome, the more closely they will be associated in gametes and the more often they will appear together; genes that are very close are essentially never separated because it is extremely unlikely that a crossover point will occur between them. This is known as genetic linkage.

History

The notion of a gene is evolving with the science of genetics, which began when Gregor Mendel noticed that biological variations are inherited from parent organisms as specific, discrete traits. The biological entity responsible for defining traits was later termed a gene, but the biological basis for inheritance remained unknown until DNA was identified as the genetic material in the 1940s. Prior to Mendel's work, the dominant theory of heredity was one of blending inheritance, which proposes that the traits of the parents blend or mix in a smooth, continuous gradient in the offspring.

Although Mendel's work was largely unrecognized after its first publication in 1866, it was rediscovered in 1900 by three European scientists, Hugo de Vries, Carl Correns, and Erich von Tschermak, who had reached similar conclusions from their own research. However, these scientists were not yet aware of the identity of the 'discrete units' on which genetic material resides.

The existence of genes was first suggested by Gregor Mendel (1822–1884), who, in the 1860s, studied inheritance in peaplants (Pisum sativum) and hypothesized a factor that conveys traits from parent to offspring. He spent over 10 years of his life on one experiment. Although he did not use the term gene, he explained his results in terms of inherited characteristics.

Mendel was also the first to hypothesize independent assortment, the distinction between dominant and recessive traits, the distinction between a heterozygote and homozygote, and the difference between what would later be described as genotype (the genetic material of an organism) and phenotype (the visible traits of that organism).

Charles Darwin used the term Gemmule to describe a microscopic unit of inheritance, and what would later become known as Chromosomes had been observed separating out during cell division byWilhelm Hofmeister as early as 1848. The idea that chromosomes are the carriers of inheritance was expressed in 1883 by Wilhelm Roux. Darwin also coined the word pangenesis by (1868). The word pangenesis is made from the Greek words pan (a prefix meaning "whole", "encompassing") and genesis ("birth") or genos ("origin").

Mendel's concept was given a name by Hugo de Vries in 1889, in his book Intracellular Pangenesis; although probably unaware of Mendel's work at the time, he coined the term "pangen" for "the smallest particle [representing] one hereditary characteristic". Danish botanist Wilhelm Johannsen coined the word "gene" ("gen" in Danish and German) in 1909 to describe the fundamental physical and functional units of heredity, while the related word genetics was first used by William Bateson in 1905. He derived the word from de Vries' "pangen". In the early 1900s, Mendel's work received renewed attention from scientists.

In 1910, Thomas Hunt Morgan showed that genes reside on specific chromosomes. He later showed that genes occupy specific locations on the chromosome. With this knowledge, Morgan and his students began the first chromosomal map of the fruit fly Drosophila. In 1928, Frederick Griffith showed that genes could be transferred. In what is now known as Griffith's experiment, injections into a mouse of a deadly strain of bacteria that had been heat-killed transferred genetic information to a safe strain of the same bacteria, killing the mouse. A series of subsequent discoveries led to the realization decades later that chromosomes within cells are the carriers of genetic material, and that they are made of DNA (deoxyribonucleic acid), apolymeric molecule found in all cells on which the 'discrete units' of Mendelian inheritance are encoded.

In 1941, George Wells Beadle and Edward Lawrie Tatum showed that mutations in genes caused errors in specific steps in metabolic pathways.

This showed that specific genes code for specific proteins, leading to the "one gene, one enzyme" hypothesis. Oswald Avery, Colin Munro MacLeod, and Maclyn McCarty showed in 1944 that DNA holds the gene's information.

In 1953, James D. Watson and Francis Crick demonstrated the molecular structure of DNA. Together, these discoveries established the central dogma of molecular biology, which states that proteins are translated from RNA which is transcribed from DNA. This dogma has since been shown to have exceptions, such as reverse transcription in retroviruses.

In 1972, Walter Fiers and his team at the Laboratory of Molecular Biology of the University of Ghent (Ghent, Belgium) were the first to determine the sequence of a gene: the gene for Bacteriophage MS2 coat protein. Richard J. Roberts and Phillip Sharp discovered in 1977 that genes can be split into segments.

This led to the idea that one gene can make several proteins. Recently (as of 2003–2006), biological results let the notion of gene appear more slippery. In particular, genes do not seem to sit side by side on DNA like discrete beads. Instead, regions of the DNA producing distinct proteins may overlap, so that the idea emerges that "genes are one long continuum".

It was first hypothesized in 1986 by Walter Gilbert that neither DNA nor protein would be required in such a primitive system as that of a very early stage of the earth if RNA could perform as simply a catalyst and genetic information storage processor.

The modern study of genetics at the level of DNA is known as molecular genetics and the synthesis of molecular genetics with traditional Darwinian evolution is known as the modern evolutionary synthesis.

Mendelian Inheritance and Classical Genetics

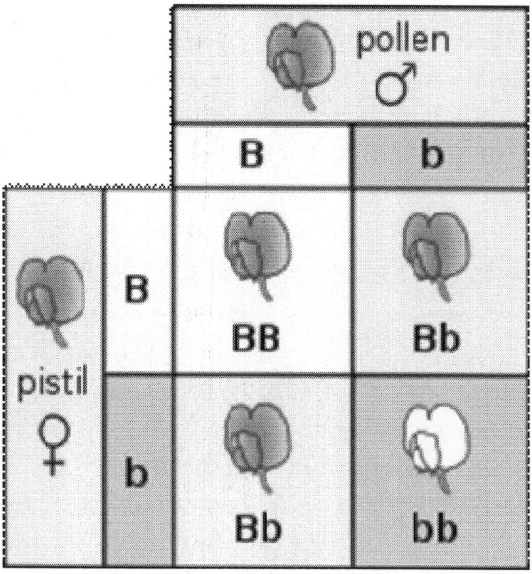

Fig. 2: Crossing between two pea plantsheterozygous for purple (B, dominant) and white (b, recessive) blossoms

According to the theory of Mendelian inheritance, variations in phenotype—the observable physical and behavioral characteristics of an organism—are due to variations in genotype, or the organism's particular set of genes, each of which specifies a particular trait. Different forms of a gene, which may give rise to different phenotypes, are known as alleles.

Organisms such as the pea plants Mendel worked on, along with many plants and animals, have two alleles for each trait, one inherited from each parent. Alleles may be dominant or recessive; dominant alleles give rise to their corresponding phenotypes when paired with any other allele for the same trait, whereas recessive alleles give rise to their corresponding phenotype only when paired with another copy of the same allele.

For example, if the allele specifying tall stems in pea plants is dominant over the allele specifying short stems, then pea plants that inherit one tall allele from one parent and one short allele from the other parent will also have tall stems.

Mendel's work found that alleles assort independently in the production of gametes, or germ cells, ensuring variation in the next generation.

Mutation

DNA replication is for the most part extremely accurate, with an error rate per site of around 10 to 10 in eukaryotes. Rare, spontaneous alterations in the base sequence of a particular gene arise from a number of sources, such as errors in DNA replication and the aftermath of DNA damage. These errors are called mutations. The cell contains many DNA repair mechanisms for preventing mutations and maintaining the integrity of the genome; however, in some cases—such as breaks in both DNA strands of a chromosome—repairing the physical damage to the molecule is a higher priority than producing an exact copy.

Due to the degeneracy of the genetic code, some mutations in protein-coding genes are silent, or produce no change in the amino acid sequence of the protein for which they code; for example, the codons UCU and UUC both code for serine, so the U"!C mutation has no effect on the protein. Mutations that do have phenotypic effects are most often neutral or deleterious to the organism, but sometimes they confer benefits to the organism's fitness.

Mutations propagated to the next generation lead to variations within a species' population. Variants of a single gene are known as alleles, and differences in alleles may give rise to differences in traits. Although it is rare for the variants in a single gene to have clearly distinguishable phenotypic effects, certain well-defined traits are in fact controlled by single genetic loci. A gene's most common allele is called the wild type allele, and rare alleles are called mutants. However, this does not imply that the wild-type allele is the ancestor from which the mutants are descended.

Genome

Chromosomal Organization

The total complement of genes in an organism or cell is known as its genome. In prokaryotes, the vast majority of genes are located on a single chromosome of circular DNA, while eukaryotes usually possess multiple individual linear DNA helices packed into dense DNA-protein complexes called chromosomes. Genes that appear together on one chromosome of one species may appear on separate chromosomes in another species. Many species carry more than one copy of their genome within each of their somatic cells. Cells or organisms with only one copy of each chromosome are called haploid;

those with two copies are called diploid; and those with more than two copies are called polyploid. The copies of genes on the chromosomes are not necessarily identical. In sexually reproducing organisms, one copy is normally inherited from each parent.

Number of Genes

The human genome, categorized by function of each gene product, given both as number of genes and as percentage of all genes.

Early estimates of the number of human genes that used expressed sequence tag data put it at 50 000–100 000. Following the sequencing of the human genome and other genomes, it has been found that rather few genes (~20 000 in human, mouse and fly, ~13 000 in roundworm, >46 000 in rice) encode all the proteins in an organism. These protein-coding sequences make up 1–2% of the human genome. A large part of the genome is transcribed however, to introns, retrotransposons and seemingly a large array of noncoding RNAs. Total number of proteins (the Earth's proteome) is estimated to be 5 million sequences.

Genetic and Genomic Nomenclature

Gene nomenclature has been established by the HUGO Gene Nomenclature Committee (HGNC) for each known human gene in the form of an approved gene name and symbol (short-form abbreviation). All approved symbols are stored in the HGNC Database. Each symbol is unique and each gene is only given one approved gene symbol. This also facilitates electronic data retrieval from publications. In preference each symbol maintains parallel construction in different members of a gene family and can be used in other species, especially themouse.

Evolutionary Concept of a Gene

George C. Williams first explicitly advocated the gene-centric view of evolution in his 1966 book Adaptation and Natural Selection. He proposed an evolutionary concept of gene to be used when we are talking about natural selection favoring some genes. The definition is: "that which segregates and recombines with appreciable frequency." According to this definition, even an asexual genome could be considered a gene, insofar that it have an appreciable permanency through many generations.

The difference is: the molecular gene transcribes as a unit, and the evolutionary gene inherits as a unit.

Richard Dawkins' books The Selfish Gene (1976) and The Extended Phenotype (1982) defended the idea that the gene is the only replicator in living systems. This means that only genes transmit their structure largely intact and are potentially immortal in the form of copies. So, genes should be the unit of selection. In The Selfish Gene Dawkins attempts to redefine the word 'gene' to mean "an inheritable unit" instead of the generally accepted definition of "a section of DNA coding for a particular protein". In River Out of Eden, Dawkins further refined the idea of gene-centric selection by describing life as a river of compatible genes flowing through geological time. Scoop up a bucket of genes from the river of genes, and we have an organism serving as temporary bodies or survival machines. A river of genes may fork into two branches representing two non-interbreeding species as a result of geographical separation.

Gene Targeting and Implications

Gene targeting is commonly referred to techniques for altering or disrupting mouse genes and provides the mouse models for studying the roles of individual genes in embryonic development, human disorders, aging and diseases. The mouse models, where one or more of its genes are deactivated or made inoperable, are called knockout mice. Since the first reports in which homologous recombination in embryonic stem cells was used to generate gene-targeted mice, gene targeting has proven to be a powerful means of precisely manipulating the mammalian genome, producing at least ten thousand mutant mouse strains and it is now possible to introduce mutations that can be activated at specific time points, or in specific cells or organs, both during development and in the adult animal.

Gene targeting strategies have been expanded to all kinds of modifications, including point mutations, isoform deletions, mutant allele correction, large pieces of chromosomal DNA insertion anddeletion, tissue specific disruption combined with spatial and temporal regulation and so on. It is predicted that the ability to generate mouse models with predictable phenotypes will have a major impact on studies of all phases of development, immunology, neurobiology, oncology, physiology, metabolism, and human diseases. Gene targeting is also in theory applicable to species from whichtotipotent embryonic stem cells can be established, and therefore may offer a potential to the improvement of domestic animals and plants.

Changing Concept

The concept of the gene has changed considerably. From the original definition of a "unit of inheritance", the term evolved to mean a DNA-based unit that can exert its effects on the organism through RNA or protein products. It was also previously believed that one gene makes one protein; this concept was overthrown by the discovery of alternative splicing and trans-splicing.

The definition of a gene is still changing. The first cases of RNA-based inheritance have been discovered in mammals. Evidence is also accumulating that the control regions of a gene do not necessarily have to be close to the coding sequence on the linear molecule or even on the same chromosome. Spilianakis and colleagues discovered that the promoter region of the interferon-gammagene on chromosome 10 and the regulatory regions of the T(H)2 cytokine locus on chromosome 11 come into close proximity in the nucleus possibly to be jointly regulated.

The concept that genes are clearly delimited is also being eroded. There is evidence for fused proteins stemming from two adjacent genes that can produce two separate protein products. While it is not clear whether these fusion proteins are functional, the phenomenon is more frequent than previously thought. Even more ground-breaking than the discovery of fused genes is the observation that some proteins can be composed of exons from far away regions and even different chromosomes. This new data has led to an updated, and probably tentative, definition of a gene as "a union of genomic sequences encoding a coherent set of potentially overlapping functional products". This new definition categorizes genes by functional products, whether they be proteins or RNA, rather than specific DNA loci; all regulatory elements of DNA are therefore classified as gene-associated regions.

Exon

An exon is a nucleic acid sequence that is represented in the mature form of an RNA molecule either after portions of a precursor RNA (introns) have been removed by cis-splicing or when two or more precursor RNA molecules have been ligated by trans-splicing. The mature RNA molecule can be a messenger RNA or a functional form of a non-coding RNA such as rRNA or tRNA. Depending on the context, exon can refer to the sequence in the DNA or its RNA transcript.

History

The term exon derives from expressed region and was coined by American biochemist Walter Gilbert in 1978: "The notion of the cistron... must be replaced by that of a transcription unit containing regions which will be lost from the mature messenger – which I suggest we call introns (for intragenic regions) – alternating with regions which will be expressed – exons."

This definition was originally made for protein-coding transcripts that are spliced before being translated. The term later came to include sequences removed from rRNA and tRNA, and it also was used later for RNA molecules originating from different parts of the genome that are then ligated by trans-splicing.

Function

In many genes, each of the exons contains part of the open reading frame (ORF) that codes for a specific portion of the complete protein. However, the term exon is often misused to refer only to coding sequences for the final protein. This is incorrect, since many noncoding exons are known in human genes (Zhang 1998).

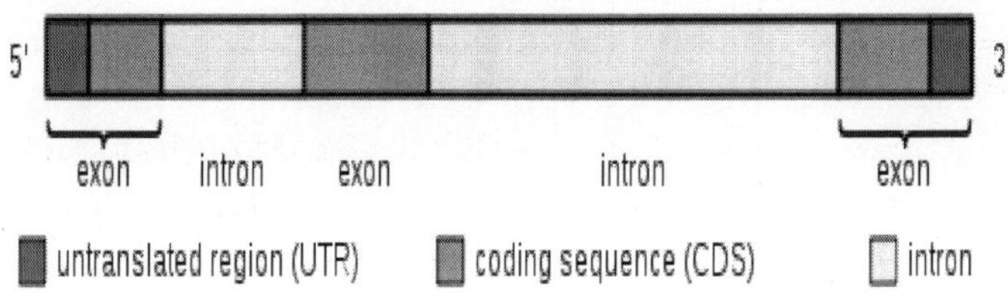

Fig. 3:

To the right is a diagram of a heterogeneous nuclear RNA (hnRNA), which is an unedited mRNA transcript, or pre-mRNAs. Exons can include both sequences that code for amino acids (red) and untranslated sequences (grey). Stretches of unused sequence called introns (blue) are removed, and the exons are joined together to form the final functional mRNA. The notation 5' and 3' refer to the direction of the DNA template in the chromosome and is used to distinguish between the two untranslated regions (grey).

Some of the exons will be wholly or part of the 5' untranslated region (5' UTR) or the 3' untranslated region (3' UTR) of each transcript. The untranslated regions are important for efficient translation of thetranscript and for controlling the rate of translation and half-life of the transcript. Furthermore, transcripts made from the same gene may not have the same exon structure, since parts of the mRNA could be removed by the process of alternative splicing. Some mRNA transcripts have exons with no ORFs and, thus, are sometimes referred to as non-coding RNA.

Exonization is the creation of a new exon, as result of mutations in intronic sequences. Polycistronic messages have multiple ORFs in one transcript and also have small regions of untranslated sequence between each ORF.

Experimental Approaches that Utilize Exons

Exon trapping or 'gene trapping' is a molecular biology technique that exploits the existence of the intron-exon splicing to find new genes. The first exon of a 'trapped' gene splices into the exon that is contained in the insertional DNA. This new exon contains the ORF for a reporter gene that can now be expressed using the enhancers that control the target gene. A scientist knows that a new gene has been trapped when the reporter gene is expressed.

Splicing can be experimentally modified so that targeted exons are excluded from mature mRNA transcripts by blocking the access of splice-directing small nuclear ribonucleoprotein particles (snRNPs) to pre-mRNA using Morpholino antisense oligos. This has become a standard technique in developmental biology. Morpholino oligos can also be targeted to prevent molecules that regulate splicing (e.g. splice enhancers, splice suppressors) from binding to pre-mRNA, altering patterns of splicing.

Intron

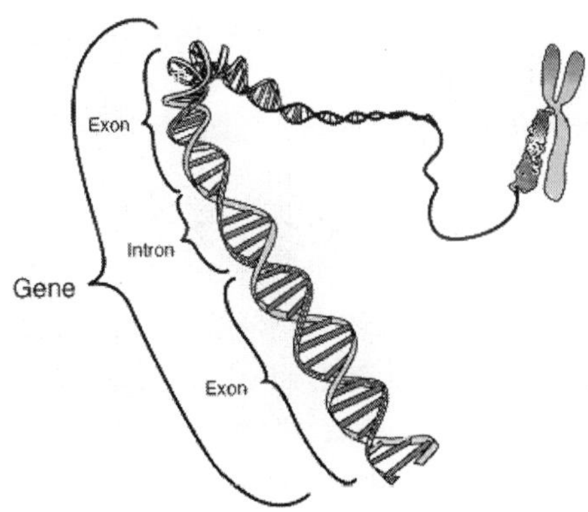

Fig. 4

An intron is any nucleotide sequence within a gene that is removed by RNA splicing to generate the final mature RNA product of a gene. The term intronrefers to both the DNA sequence within a gene, and the corresponding sequence in RNA transcripts. Sequences that are joined together in the final mature RNA after RNA splicing are exons. Introns are found in the genes of most organisms and many viruses, and can be located in a wide range of genes, including those that generate proteins, ribosomal RNA (rRNA), and transfer RNA (tRNA). When proteins are generated from intron-containing genes, RNA splicing takes place as part of the RNA processing pathway that follows transcription and precedes translation.

The word intron is derived from the term intragenic region, i.e. a region inside a gene. Although introns are sometimes called intervening sequences, the term "intervening sequence" can refer to any of several families of internal nucleic acid sequences that are not present in the final gene product, including inteins, untranslated sequences (UTR), and nucleotides removed by RNA editing, in addition to introns.

Introduction

Introns were first discovered in protein-coding genes of adenovirus, and were subsequently identified in genes encoding transfer RNA and ribosomal RNA genes. Introns are now known to occur within a wide variety of genes throughout organisms and viruses within all of the biological kingdoms.

The discovery of introns led to the Nobel Prize in Physiology or Medicine in 1993 for Phillip Allen Sharp and Richard J. Roberts. The term intron was introduced by American biochemist Walter Gilbert:

"The notion of the cistron [...] must be replaced by that of a transcription unit containing regions which will be lost from the mature messenger - which I suggest we call introns (for intragenic regions) - alternating with regions which will be expressed - exons." (Gilbert 1978)

The frequency of introns within different genomes is observed to vary widely across the spectrum of biological organisms. For example, introns are extremely common within the nuclear genome of higher vertebrates (e.g. humans and mice), where protein-coding genes almost always contain multiple introns, while introns are rare within the nuclear genes of some eukaryotic microorganisms, for example baker's yeast (Saccharomyces cerevisiae). In contrast, the mitochondrial genomes of vertebrates are entirely devoid of introns, while those of eukaryotic microorganisms may contain many introns. Introns are well known in bacterial and archaeal genes, but occur more rarely than in most eukaryotic genomes.

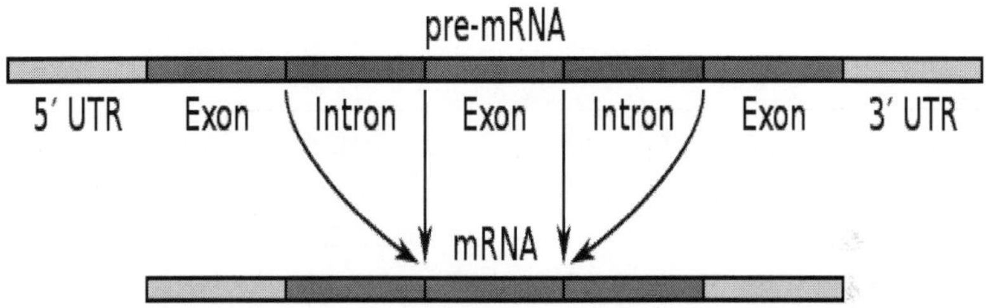

Fig. 5:

Simple illustration of an unspliced mRNA precursor, with two introns and three exons (top). After the introns have been removed via splicing, the mature mRNA sequence is ready for translation (bottom).

Classification

Splicing of all intron-containing RNA molecules is superficially similar, as described above. However, different types of introns were identified through the examination of intron structure by DNA sequence analysis, together with genetic and biochemical analysis of RNA splicing reactions.

At least four distinct classes of introns have been identified.

§ Introns in nuclear protein-coding genes that are removed by spliceosomes

§ Introns in nuclear and archaeal transfer RNA genes that are removed by proteins (tRNA splicing enzymes)

§ Self-splicing group I introns that are removed by RNA catalysis.

§ Self-splicing group II introns that are removed by RNA catalysis

Group III introns are proposed to be a fifth family, but little is known about the biochemical apparatus that mediates their splicing. They appear to be related to group II introns, and possibly to spliceosomal introns.

Nuclear pre-mRNA introns (spliceosomal introns) are characterized by specific intron sequences located at the boundaries between introns and exons. These sequences are recognized by spliceosomal RNA molecules when the splicing reactions are initiated. In addition, they contain a branch point, a particular nucleotide sequence near the 3' end of the intron that becomes covalently linked to the 5' end of the intron during the splicing process, generating a branched (lariat) intron. Apart from these three short conserved elements, nuclear pre-mRNA intron sequences are highly variable. Nuclear pre-mRNA introns are often much longer than their surrounding exons.

Group I and group II introns are found in genes encoding proteins (messenger RNA), transfer RNA and ribosomal RNA in a very wide range of living organisms., Following transcription into RNA, group I and group II introns also make extensive internal interactions that allow them to fold into a specific, complex three-dimensional architecture. These complex architectures allow some group I and group II introns to be self-splicing, that is, the intron-containing RNA molecule can rearrange its own covalent structure so as to precisely remove the intron and link the exons together in the correct order. In some cases, particular intron-binding proteins are involved in splicing, acting in such a way that the assist the intron in folding into the three-dimensional structure that is necessary for self-splicing activity. Group I and group II introns are distinguished by different sets of internal conserved sequences and folded structures, and by the fact that splicing of RNA molecules containing group II introns generates branched introns (like those of spliceosomal RNAs), while group I introns use a non-encoded guanosine nucleotide (typically GTP) to initiate splicing, adding it on to the 5'-end of the excised intron.

Transfer RNA introns that depend upon proteins for removal occur at a specific location within the anticodon loop of unspliced tRNA precursors, and are removed by a tRNA splicing endonuclease. The exons are then linked together by a second protein, the tRNA splicing ligase. Note that self-splicing introns are also sometimes found within tRNA genes.

Biological Functions and Evolution

As a first approximation, it is possible to view introns as unimportant sequences whose only function is to be removed from an unspliced precursor RNA in order to generate the functional mRNA, rRNA or tRNA product. However, it is now well-established that some introns themselves encode specific proteins or can be further processed after splicing to generate noncoding RNA molecules.Alternative splicing is widely used to generate multiple proteins from a single gene. Furthermore, some introns represent mobile genetic elements and may be regarded as examples of selfish DNA.

The biological origins of introns are obscure. After the initial discovery of introns in protein-coding genes of the eukaryotic nucleus, there was significant debate as to whether introns in modern-day organisms were inherited from a common ancient ancestor (termed the introns-early hypothesis), or whether they appeared in genes rather recently in the evolutionary process (termed the introns-late hypothesis). Another theory is that the spliceosome and the intron-exon structure of genes is a relic of the RNA world (the introns-first hypothesis). There is still considerable debate about the extent to which of these hypotheses is most correct. The popular consensus at the moment is that introns arose within the eukaryote lineage as selfish elements.

Early studies of genomic DNA sequences from a wide range of organisms show that the intron-exon structure of homologous genes in different organisms can vary widely. More recent studies of entire eukaryotic genomes have now shown that the lengths and density (introns/gene) of introns varies considerably between related species. For example, while the human genome contains an average of 8.4 introns/gene (139,418 in the genome), thc unicellular fungus Encephalitozoon cuniculi contains only 0.0075 introns/gene (15 introns in the genome).

Since eukaryotes arose from a common ancestor (Common descent), there must have been extensive gain and/or loss of introns during evolutionary time. This process is thought to be subject to selection, with a tendency towards intron gain in larger species due to their smaller population sizes, and the converse in smaller (particularly unicellular) species. Biological factors also influence which genes in a genome lose or accumulate introns.

Alternative splicing of introns within a gene acts to introduce greater variability of protein sequences translated from a single gene, allowing multiple related proteins to be generated from a single gene and a single precursor mRNA transcript. The control of alternative RNA splicing is performed by complex network of signaling molecules that respond to a wide range of intracellular and extracellular signals.

Introns contain several short sequences that are important for efficient splicing, such as acceptor and donor sites at either end of the intron as well as a branch point site, which are required for proper splicing by the spliceosome. Some introns are known to enhance the expression of the gene that they are contained in by a process known as intron-mediated enhancement (IME).

Description

Mutations can involve large sections of DNA becoming duplicated, usually through genetic recombination. These duplications are a major source of raw material for evolving new genes, with tens to hundreds of genes duplicated in animal genomes every million years. Most genes belong to larger families of genes of shared ancestry. Novel genes are produced by several methods, commonly through the duplication and mutation of an ancestral gene, or by recombining parts of different genes to form new combinations with new functions.

Here, domains act as modules, each with a particular and independent function, that can be mixed together to produce genes encoding new proteins with novel properties. For example, the human eye uses four genes to make structures that sense light: three for color vision and one for night vision; all four arose from a single ancestral gene. Another advantage of duplicating a gene (or even anentire genome) is that this increases redundancy; this allows one gene in the pair to acquire a new function while the other copy performs the original function. Other types of mutation occasionally create new genes from previously noncoding DNA.

Changes in chromosome number may involve even larger mutations, where segments of the DNA within chromosomes break and then rearrange. For example, in the Homininae, two chromosomes fused to produce human chromosome 2; this fusion did not occur in the lineage of the other apes, and they retain these separate chromosomes. In evolution, the most important role of such chromosomal rearrangements may be to accelerate the divergence of a population into new species by making populations less likely to interbreed, and thereby preserving genetic differences between these populations.

Sequences of DNA that can move about the genome, such as transposons, make up a major fraction of the genetic material of plants and animals, and may have been important in the evolution of genomes. For example, more than a million copies of the Alu sequence are present in the human genome, and these sequences have now been recruited to perform functions such as regulatinggene expression. Another effect of these mobile DNA sequences is that when they move within a genome, they can mutate or delete existing genes and thereby produce genetic diversity.

Nonlethal mutations accumulate within the gene pool and increase the amount of genetic variation. The abundance of some genetic changes within the gene pool can be reduced by natural selection, while other "more favorable" mutations may accumulate and result in adaptive changes.

For example, a butterfly may produce offspring with new mutations. The majority of these mutations will have no effect; but one might change the color of one of the butterfly's offspring, making it harder (or easier) for predators. If this color change is advantageous, the chance of this butterfly surviving and producing its own offspring are a little better, and over time the number of butterflies with this mutation may form a larger percentage of the population.

Neutral mutations are defined as mutations whose effects do not influence the fitness of an individual. These can accumulate over time due to genetic drift. It is believed that the overwhelming majority of mutations have no significant effect on an organism's fitness. Also, DNA repair mechanisms are able to mend most changes before they become permanent mutations, and many organisms have mechanisms for eliminating otherwise permanently mutated somatic cells.

Mutation is generally accepted by biologists as the mechanism by which natural selection acts, generating advantageous new traits that survive and multiply in offspring as well as disadvantageous traits, in less fit offspring, that tend to die out.

Causes

Two classes of mutations are spontaneous mutations (molecular decay) and induced mutations caused by mutagens.

Spontaneous Mutation

Spontaneous mutations on the molecular level can be caused by:

§ Tautomerism – A base is changed by the repositioning of a hydrogen atom, altering the hydrogen bonding pattern of that base resulting in incorrect base pairing during replication.

- § Depurination – Loss of a purine base (A or G) to form an apurinic site (AP site).
- § Deamination – Hydrolysis changes a normal base to an atypical base containing a keto group in place of the original amine group. Examples include C '! U and A '! HX (hypoxanthine), which can be
- § corrected by DNA repair mechanisms; and 5MeC (5-methylcytosine) '! T, which is less likely to be detected as a mutation because thymine is a normal DNA base.
- § Slipped strand mispairing - Denaturation of the new strand from the template during replication, followed by renaturation in a different spot ("slipping"). This can lead to insertions or deletions.

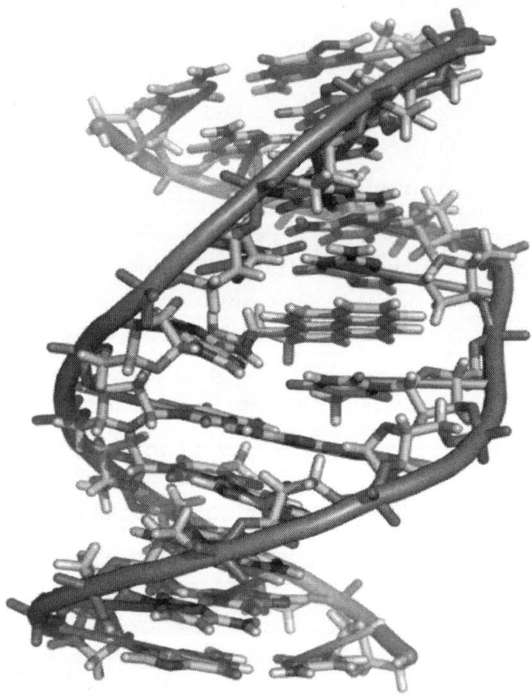

Fig. 6: A covalent adduct between benzo[a]pyrene,the major mutagen in tobacco smoke, and DNA

Induced Mutation

Induced mutations on the molecular level can be caused by:

- § Chemicals

§ Hydroxylamine NH_2OH

§ Base analogs (e.g. BrdU)

§ Alkylating agents (e.g. N-ethyl-N-nitrosourea) These agents can mutate both replicating and non-replicating DNA. In contrast, a base analog can only mutate the DNA when the analog is incorporated in replicating the DNA. Each of these classes of chemical mutagens has certain effects that then lead to transitions, transversions, or deletions.

§ Agents that form DNA adducts (e.g. ochratoxin A metabolites)

§ DNA intercalating agents (e.g. ethidium bromide)

§ DNA crosslinkers

§ Oxidative damage

§ Nitrous acid converts amine groups on A and C to diazo groups, altering their hydrogen bonding patterns which leads to incorrect base pairing during replication.

§ Radiation

§ Ultraviolet radiation (nonionizing radiation). Two nucleotide bases in DNA – cytosine and thymine – are most vulnerable to radiation that can change their properties. UV light can induce adjacent pyrimidine bases in a DNA strand to become covalently joined as a pyrimidine dimer. UV radiation, particularly longer-wave UVA, can also cause oxidative damage to DNA.

§ Ionizing radiation

§ Radioactive decay, such as C in DNA

§ Viral infections

DNA has so-called hotspots, where mutations occur up to 100 times more frequently than the normal mutation rate. A hotspot can be at an unusual base, e.g., 5-methylcytosine.

Mutation rates also vary across species. Evolutionary biologists have theorized that higher mutation rates are beneficial in some situations, because they allow organisms to evolve and therefore adapt more quickly to their environments. For example, repeated exposure of bacteria to antibiotics, and selection of resistant mutants, can result in the selection of bacteria that have a much higher mutation rate than the original population (mutator strains).

Classification of Mutation Types

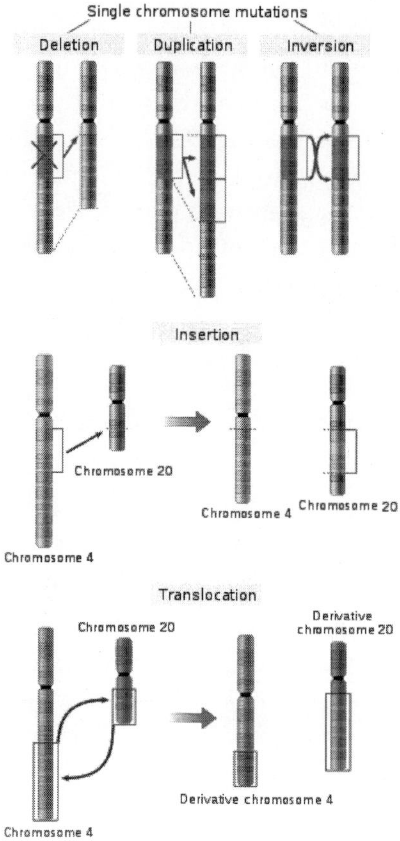

Fig. 7: Illustrations of five types of chromosomal mutations.

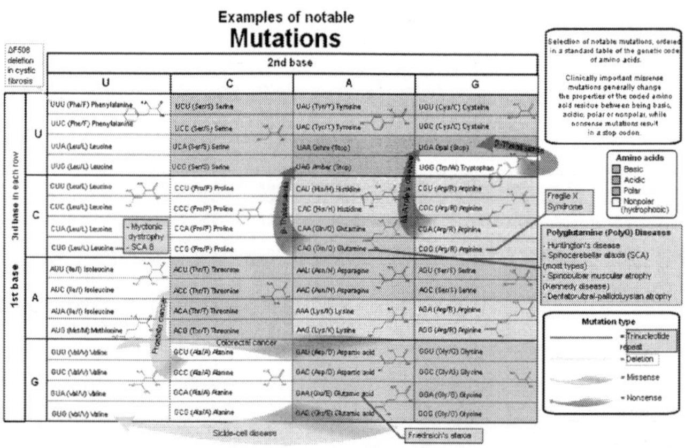

Fig. 8: Selection of disease-causing mutations, in a standard table of the genetic code of amino acids.

By Effect on Structure

The sequence of a gene can be altered in a number of ways. Gene mutations have varying effects on health depending on where they occur and whether they alter the function of essential proteins. Mutations in the structure of genes can be classified as:

§ Small-scale mutations, such as those affecting a small gene in one or a few nucleotides, including:

§ Point mutations, often caused by chemicals or malfunction of DNA replication, exchange a single nucleotide for another. These changes are classified as transitions or transversions. Most common is the transition that exchanges a purine for a purine (A "! G) or a pyrimidine for a pyrimidine, (C "! T). A transition can be caused by nitrous acid, base mis-pairing, or mutagenic base analogs such as5-bromo-2-deoxyuridine (BrdU). Less common is a transversion, which exchanges a purine for a pyrimidine or a pyrimidine for a purine (C/T "! A/G). An example of a transversion is adenine (A) being converted into a cytosine (C). A point mutation can be reversed by another point mutation, in which the nucleotide is changed back to its original state (true reversion) or by second-site reversion (a complementary mutation elsewhere that results in regained gene functionality). Point mutations that occur within the protein coding region of a gene may be classified into three kinds, depending upon what the erroneous codon codes for:

§ Silent mutations: which code for the same amino acid.

§ Missense mutations: which code for a different amino acid.

§ Nonsense mutations: which code for a stop and can truncate the protein.

§ Insertions add one or more extra nucleotides into the DNA. They are usually caused by transposable elements, or errors during replication of repeating elements (e.g. AT repeats). Insertions in the coding region of a gene may alter splicing of the mRNA(splice site mutation), or cause a shift in the reading frame (frameshift), both of which can significantly alter the gene product. Insertions can be reverted by excision of the transposable element.

§ Deletions remove one or more nucleotides from the DNA. Like insertions, these mutations can alter the reading frame of the gene. They are generally irreversible: though exactly the same sequence might theoretically be restored by an insertion, transposable elements able to revert a very short deletion (say 1–2 bases) in any location are either highly unlikely t o exist or do not exist at all. Note that a deletion is not the exact opposite of an insertion: the former is quite random while the latter consists of a specific sequence inserting at locations that are not entirely random or even quite narrowly defined.

§ Large-scale mutations in chromosomal structure, including:

§ Amplifications (or gene duplications) leading to multiple copies of all chromosomal regions, increasing the dosage of the genes located within them.

§ Deletions of large chromosomal regions, leading to loss of the genes within those regions.

§ Mutations whose effect is to juxtapose previously separate pieces of DNA, potentially bringing together separate genes to form functionally distinct fusion genes (e.g. bcr-abl). These include:

§ Chromosomal translocations: interchange of genetic parts from nonhomologous chromosomes.

§ Interstitial deletions: an intra-chromosomal deletion that removes a segment of DNA from a single chromosome, thereby apposing previously distant genes. For example, cells isolated from a human astrocytoma, a type of brain tumor, were found to have a chromosomal deletion removing sequences between the "fused in glioblastoma"gene and the receptor tyrosine kinase "ros", producing a fusion protein. The abnormal FIG-ROS fusion protein has constitutively active kinase activity that causes oncogenic transformation (a transformation from normal cells to cancer cells).

§ Chromosomal inversions: reversing the orientation of a chromosomal segment.

§ Loss of heterozygosity: loss of one allele, either by a deletion or recombination event, in an organism that previously had two different alleles.

By Effect on Function

- § Loss-of-function mutations are the result of gene product having less or no function. When the allele has a complete loss of function (null allele) it is often called an amorphic mutation. Phenotypes associated with such mutations are most often recessive. Exceptions are when the organism is haploid, or when the reduced dosage of a normal gene product is not enough for a normal phenotype (this is calledhaploinsufficiency).
- § Gain-of-function mutations change the gene product such that it gains a new and abnormal function. These mutations usually have dominant phenotypes. Often called a neomorphic mutation.
- § Dominant negative mutations (also called antimorphic mutations) have an altered gene product that acts antagonistically to the wild-type allele. These mutations usually result in an altered molecular function (often inactive) and are characterised by a dominant or semi-dominant phenotype. In humans, Marfan syndrome is an example of a dominant negative mutation occurring in anautosomal dominant disease. In this condition, the defective glycoprotein product of the fibrillin gene (FBN1) antagonizes the product of the normal allele.
- § Lethal mutations are mutations that lead to the death of the organisms which carry the mutations.
- § A back mutation or reversion is a point mutation that restores the original sequence and hence the original phenotype.

By Effect on Fitness

In applied genetics it is usual to speak of mutations as either harmful or beneficial.

- § A harmful mutation is a mutation that decreases the fitness of the organism.
- § A beneficial mutation is a mutation that increases fitness of the organism, or which promotes traits that are desirable.

  In theoretical population genetics, it is more usual to speak of such mutations as deleterious or advantageous. In the neutral theory of molecular evolution, genetic drift is the basis for most variation at the molecular level.

§ A neutral mutation has no harmful or beneficial effect on the organism. Such mutations occur at a steady rate, forming the basis for the molecular clock.

§ A deleterious mutation has a negative effect on the phenotype, and thus decreases the fitness of the organism.

§ An advantageous mutation has a positive effect on the phenotype, and thus increases the fitness of the organism.

§ A nearly neutral mutation is a mutation that may be slightly deleterious or advantageous, although most nearly neutral mutations are slightly deleterious.

In reality, viewing the fitness effects of mutations in these discrete categories is an oversimplification. Attempts have been made to infer the distribution of fitness effects using mutagenesis experiments or theoretical models applied to molecular sequence data. However, the current distribution is still uncertain, and some aspects of the distribution likely vary between species.

By Inheritance

§ inheritable generic in pro-generic tissue or cells on path to be changed to gametes.

§ non inheritable somatic (e.g., carcinogenic mutation)

§ non inheritable post mortem aDNA mutation in decaying remains.

By pattern of inheritance The human genome contains two copies of each gene – a paternal and a maternal allele.

§ A heterozygous mutation is a mutation of only one allele.

§ A homozygous mutation is an identical mutation of both the paternal and maternal alleles.

§ Compound heterozygous mutations or a genetic compound comprises two different mutations in the paternal and maternal alleles.

§ A wildtype or homozygous non-mutated organism is one in which neither allele is mutated. (Just not a mutation)

By Impact on Protein Sequence

§ A frameshift mutation is a mutation caused by insertion or deletion of a number of nucleotides that is not evenly divisible by three from a DNA sequence. Due to the triplet nature of gene expression by codons, the insertion or deletion can disrupt the reading frame, or the grouping of the codons, resulting in a completely different translation from the original. The earlier in the sequence the deletion or insertion occurs, the more altered the protein produced is.

In contrast, any insertion or deletion that is evenly divisible by three is termed an in-frame mutation

§ A nonsense mutation is a point mutation in a sequence of DNA that results in a premature stop codon, or a nonsense codon in the transcribed mRNA, and possibly a truncated, and often nonfunctional protein product.

§ Missense mutations or nonsynonymous mutations are types of point mutations where a single nucleotide is changed to cause substitution of a different amino acid. This in turn can render the resulting protein nonfunctional. Such mutations are responsible for diseases such as Epidermolysis bullosa, sickle-cell disease, and SOD1 mediated ALS (Boillée 2006, p. 39).

§ A neutral mutation is a mutation that occurs in an amino acid codon which results in the use of a different, but chemically similar, amino acid. The similarity between the two is enough that little or no change is often rendered in the protein. For example, a change from AAA to AGA will encode arginine, a chemically similar molecule to the intended lysine. Neutral mutations occur because of the degenerate nature of the genetic code.

§ Silent mutations are mutations that do not result in a change to the amino acid sequence of a protein. They may occur in a region that does not code for a protein, or they may occur within a codon in a manner that does not alter the final amino acid sequence. The phrase silent mutation is often used interchangeably with the phrase synonymous mutation; however, synonymous mutations are a subcategory of the former, occurring only within exons. The name silent could be a misnomer. For example, a silent mutation in the exon/intron border may lead to alternative splicing by changing the splice site, thereby leading to a changed protein.

By Inheritance Ability

Fig. 9:

A mutation has caused this garden moss rose to produce flowers of different colors. This is a somatic mutation that may also be passed on in the germ line.

In multicellular organisms with dedicated reproductive cells, mutations can be subdivided into germ line mutations, which can be passed on to descendants through their reproductive cells, and somatic mutations (also called acquired mutations), which involve cells outside the dedicated reproductive group and which are not usually transmitted to descendants. A germline mutation gives rise to a constitutional mutation in the offspring, that is, a mutation that is present in every cell. A constitutional mutation can also occur very soon after fertilisation, or continue from a previous constitutional mutation in a parent.

If the organism can reproduce asexually through mechanisms such as cuttings or budding the distinction can become blurred. For example, plants can sometimes transmit somatic mutations to their descendants asexually or sexually where flower buds develop in somatically mutated parts of plants. A new mutation that was not inherited from either parent is called a de novo mutation. The source of the mutation is unrelated to the consequence, although the consequences are related to which cells were mutated.

Special Classes

- § Conditional mutation is a mutation that has wild-type (or less severe) phenotype under certain “permissive” environmental conditions and a mutant phenotype under certain “restrictive” conditions. For example, a temperature-sensitive mutation can cause cell death at high temperature (restrictive condition), but might have no deleterious consequences at a lower temperature (permissive condition).

Nomenclature

A committee of the Human Genome Variation Society (HGVS) has developed the standard human sequence variant nomenclature, which should be used by researchers and DNA diagnostic centers to generate unambiguous mutation descriptions. In principle, this nomenclature can also be used to describe mutations in other organisms. The nomenclature specifies the type of mutation and base or amino acid changes.

- § Nucleotide substitution (e.g. 76A>T) - The number is the position of the nucleotide from the 5' end, the first letter represents the wild type nucleotide, and the second letter represents the nucleotide which replaced the wild type. In the given example, the adenine at the 76th position was replaced by a thymine.
- § If it becomes necessary to differentiate between mutations in genomic DNA, mitochondrial DNA, and RNA, a simple convention is used. For example, if the 100th base of a nucleotide sequence mutated from G to C, then it would be written as g.100G>C if the mutation occurred in genomic DNA, m.100G>C if the mutation occurred in mitochondrial DNA, or r.100g>c if the mutation occurred in RNA. Note that for mutations in RNA, the nucleotide code is written in lower case.
- § Amino acid substitution (e.g. D111E) – The first letter is the one letter code of the wild type amino acid, the number is the position of the amino acid from the N-terminus, and the second letter is the one letter code of the amino acid present in the mutation. Nonsense mutations are represented with an X for the second amino acid (e.g. D111X).
- § Amino acid deletion (e.g. ÄF508) – The Greek letter Ä (delta) indicates a deletion. The letter refers to the amino acid present in the wild type and the number is the position from the N

terminus of the amino acid were it to be present as in the wild type.

Harmful Mutations

Changes in DNA caused by mutation can cause errors in protein sequence, creating partially or completely non-functional proteins. To function correctly, each cell depends on thousands of proteins to function in the right places at the right times. When a mutation alters a protein that plays a critical role in the body, a medical condition can result. A condition caused by mutations in one or more genes is called a genetic disorder. Some mutations alter a gene's DNA base sequence but do not change the function of the protein made by the gene. Studies of the fly Drosophila melanogastersuggest that if a mutation does change a protein, this will probably be harmful, with about 70 percent of these mutations having damaging effects, and the remainder being either neutral or weakly beneficial. However, studies in yeast have shown that only 7% of mutations that are not in genes are harmful.

If a mutation is present in a germ cell, it can give rise to offspring that carries the mutation in all of its cells. This is the case in hereditary diseases. On the other hand, a mutation may occur in asomatic cell of an organism. Such mutations will be present in all descendants of this cell within the same organism, and certain mutations can cause the cell to become malignant, and thus causecancer.

Often, gene mutations that could cause a genetic disorder are repaired by the DNA repair system of the cell. Each cell has a number of pathways through which enzymes recognize and repair mistakes in DNA. Because DNA can be damaged or mutated in many ways, the process of DNA repair is an important way in which the body protects itself from disease.

Beneficial Mutations

Although mutations that change protein sequences are predominantly harmful to an organism; on occasion, the effect can be neutral or positive in a given environment. In this case, the mutation may enable the mutant organism to withstand particular environmental stresses better than wild-type organisms, or reproduce more quickly. In these cases a mutation will tend to become more common in a population through natural selection.

For example, a specific 32 base pair deletion in human CCR5 (CCR5-Ä32) confers HIV resistance to homozygotes and delays AIDS onset in heterozygotes. The CCR5 mutation is more common in those of European descent. One possible explanation of the etiology of the relatively high frequency of CCR5-Ä32 in the European population is that it conferred resistance to the bubonic plague in mid-14th century Europe. People with this mutation were more likely to survive infection; thus its frequency in the population increased. This theory could explain why this mutation is not found in southern Africa, where the bubonic plague never reached. A newer theory suggests that the selective pressure on the CCR5 Delta 32 mutation was caused by smallpox instead of the bubonic plague.

Another example is Sickle cell disease, a blood disorder in which the body produces an abnormal type of the oxygen-carrying substance hemoglobin in the red blood cells. One-third of all indigenousinhabitants of Sub-Saharan Africa carry the gene, because in areas where malaria is common, there is a survival value in carrying only a single sickle-cell gene (sickle cell trait). Those with only one of the two alleles of the sickle-cell disease are more resistant to malaria, since the infestation of the malaria plasmodium is halted by the sickling of the cells which it infests.

Prion Mutation

Prions are proteins and do not contain genetic material. However, prion replication has been shown to be subject to mutation and natural selection just like other forms of replication.

Mutagen

In genetics, a mutagen (Latin, literally origin of change) is a physical or chemical agent that changes the genetic material, usually DNA, of an organism and thus increases the frequency of mutationsabove the natural background level. As many mutations cause cancer, mutagens are therefore also likely to be carcinogens. Not all mutations are caused by mutagens: so-called “spontaneous mutations” occur due to errors in DNA replication, repair and recombination.

Discovery of Mutagens

The first mutagens to be identified were carcinogens, substances that were shown to be linked to cancer. Tumors were described more than 2,000 years before the discovery of chromosomes and DNA; in

500 B.C., the Greek physician Hippocrates named crab-shaped tumors karkinos (from which the word "cancer" is derived via Latin), meaning crab. In 1567, Swiss physician Paracelsus suggested unidentified substance in mined ore caused a wasting disease in miners, and in England, in 1761, John Hill made the first direct link of cancer to chemical substances by noting that excessive use ofsnuff may cause nasal cancer. In 1775, Dr. Percivall Pott wrote a paper on the high incidence of scrotal cancer in chimney sweeps, and suggested chimney soot as the cause of scrotal cancer. In 1915, Yamagawa and Ichikawa showed that repeated application of coal tar to rabbit's ears produced malignant cancer. Subsequently in the 1930s the carcinogen component in coal tar was identified as a polyaromatic hydrocarbon (PAH), benzo(a)pyrene. Polyaromatic hydrocarbons are also present in soot which was suggested to be a causative agent of cancer over 150 years earlier.

The mutagenic property of mutagens was first demonstrated in 1927, when Hermann Muller discovered that x-rays can cause genetic mutations in fruit flies, producing phenotypic mutants as well as observable changes to the chromosomes. His collaborator Edgar Altenburg also demonstrated the mutational effect of UV radiation in 1928. Muller went on to use x-rays to create Drosophilamutants that he used in his studies of genetics. He also found that x-rays not only mutate genes in fruit flies but also have effects on the genetic makeup of humans. Similar work by Lewis Stadleralso showed the mutational effect of X-ray on barley in 1928, and ultraviolet (UV) radiation on maize in 1936. The effect of sunlight had previously been noted in the nineteenth century where rural outdoor workers and sailors were more prone to skin cancer.

Chemical mutagens were not demonstrated to cause mutation until the 1940s, when Charlotte Auerbach and J. M. Robson, found that mustard gas can cause mutations in fruit flies. A large number of chemical mutagens have since been identified, especially after the development of the Ames test in 1970s by Bruce Ames that screens for mutagens and allows for preliminary identification of carcinogens. Early studies by Ames showed around 90% of known carcinogens can be identified in Ames test as mutagenic, and 80% of the mutagens identified through Ames test are also carcinogens. Mutagens are not necessarily carcinogens, and vice versa. Sodium Azide for example may be mutagenic (and highly toxic), but it has not been shown to be carcinogenic.

Effects of Mutagens

Mutagen causes changes to the DNA which can affect the transcription and replication of the DNA, which in severe cases can lead to cell death. The mutagen produces mutations in the DNA, and deleterious mutation can result in aberrant, impaired or loss of function for a particular gene, and accumulation of mutations may lead to cancer. Different mutagens act on the DNA differently. Powerful mutagens may result in chromosomal instability, causing chromosomal breakages and rearrangement of the chromosomes such astranslocation, deletion, and inversion. Such mutagens are called clastogens.

Mutagens may also modified the DNA sequence, the changes in nucleic acid sequences by mutations include substitution of nucleotide base-pairs and insertions and deletions of one or more nucleotides in DNA sequences. Although some of these mutations are lethal or can cause serious disease, many have minor effects as they do not result in residue changes that have significant effect on the structure and function of the proteins. Many mutations are silent mutations, causing no visible effects at all, either because they occur in non-coding or non-functional sequences, or they do not change the amino-acid sequence due to the redundancy of codons.

Some mutagens can cause aneuploidy and polyploidy due to loss or gain of one or more chromosomes. In Ames test, where the varying concentrations of the chemical are used in the test, the dose response curve obtained is nearly always linear, suggesting that there is no threshold for mutagenesis. Similar results are also obtained in studies with radiations, indicating that there may be no safe threshold for mutagens. However, some proposed that low level of some mutagens may stimulate the DNA repair processes and therefore may not necessarily be harmful.

Types of Mutagens

Mutagens may be of physical, chemical or biological origin. They may act directly on the DNA, causing direct damage to the DNA, and most often result in replication error. Some however may act on the replication mechanism and chromosomal partition. Many mutagens are not mutagenic by themselves, but can form mutagenic metabolites through cellular processes. Such mutagens are called promutagens.

Chapter 5

Physical Mutagens

- § Ionizing radiations such as X-rays, gamma rays and alpha particles may cause DNA breakage and other damages.
- § Ultraviolet radiations with wavelength above 260 nm are absorbed strongly by bases, producing pyrimidine dimers which can cause error in replication if left uncorrected.
- § Radioactive decay, such as C in DNA.

DNA Reactive Chemicals

A large number of chemicals may interact directly with DNA. However, many such as PAHs, aromatic amines, benzene are not necessarily mutagenic by themselves, but through metabolic processes in cells they produce mutagenic compounds.

- § Reactive oxygen species (ROS) - These may be superoxide, hydroxyl radicals and hydrogen peroxide, and large number of these highly reactive species are generated by normal cellular processes, for example as a by-products of mitochondrial electron transport, or lipid peroxidation. A number of mutagens may also generate these ROS.
- § Deaminating agents such as nitrous acid
- § Polycyclic aromatic hydrocarbon (PAH)
- § Alkylating agents such as ethylnitrosourea. The compounds transfer methyl or ethyl group to bases or the backbone phosphate groups. Guanine when alkylated may be mispaired with thiamine. Some may cause DNA crosslinking and breakages. Nitrosamines are an important group of mutagens found in tobacco, and may also be formed in in smoked meats and fish via the interaction of amines in food with nitrites added as preservatives. Other alkylating agents include mustard gas and vinyl chloride.

§ Aromatic aminess and amides have been associated with carcinogenesis since 1895 when German physician Ludwig Rehn observed high incidence of bladder cancer among workers in German synthetic aromatic amine dye industry. 2-Acetylaminofluorene, originally used as a pesticide but may also be found in cooked meat, may cause cancer of the bladder, liver, ear, intestine, thyroid and breast.

§ Alkaloid from plants, such as those from Vinca species, may be converted by metabolic processes into the active mutagen or carcinogen

§ Bromine and some compounds that contain bromine in their chemical structure

§ Sodium azide, an azide salt that is a common reagent in organic synthesis and a component in many car airbag systems

§ Psoralen combined with ultraviolet radiation causes DNA cross-linking and hence chromosome breakage

§ Benzene, an industrial solvent and precursor in the production of drugs, plastics, synthetic rubber and dyes.

Base Analogs

§ Base analog, which can substitute for DNA bases and cause copying errors.

Intercalating Agents

§ Intercalating agents such as ethidium bromide and proflavine are molecules that may insert between bases in DNA, causing frameshift mutation during replication. Some such as daunorubicin may block transcription and replication.

Metals

Many metals, such as Arsenic, cadmium, chromium, nickel and their compounds may be mutagenic, they may however act via different mechanisms. Arsenic, chromium, iron, and nickel may be associated with the production of reactive oxygen species, nickel may also be linked to DNA hypermethylation and histone deacetylation, while cadmium may inhibit DNA mismatch repair.

Biological Agents

§ Transposon, a section of DNA that undergoes autonomous fragment relocation/multiplication. Its insertion into chromosomal DNA disrupt functional elements of the genes.

- § Virus - Virus DNA may be inserted into the genome and disrupts genetic function.
- § Bacteria - some bacteria such as Helicobacter pylori cause inflammation during which oxidative species are produced, causing DNA damage by reducing efficiency of DNA repair systems thereby increasing mutation.

Protection Against Mutagens

Antioxidants are important groups of compounds that may help remove ROS. These may be found in fruit and vegetables.

Limit exposure to mutagens such as UV radiation and tobacco smoke, control of infectious agents known to cause cancer.

Mutagen Test Systems

Many different systems for detecting mutagen have been developed. Animal systems may more accurately reflect the metabolism of human, however, they are expensive and time-consuming (may take around three years to complete), they are therefore not used as a first screen for mutagenicity or carcinogenicity.

Bacterial Systems

- § Ames test - This is the most commonly used test using Salmonella typhimurium strains deficient in histidine biosynthesis. The test checks for mutants that can be reverted back to wild-type. It is an easy, inexpensive and convenient initial screen for mutagens.
- § Resistance to 8-azaguanine in S. typhimurium - Similar to Ames test, but instead of reverse mutation, it checks for forward mutation that confer resistance to 8-azaguanine in a histidine revertant strain.
- § Escherichia coli systems - Both forward and reverse mutation detection system have been modified for use in E. coli. Tryptophane-deficient mutant is used for the reverse mutation, while galactose utility or resistance to 5-methyltryptophane may be used for forward mutation.
- § DNA repair - E. coli and Bacillus subtilis strains deficient in DNA repair may be used to detect mutagens by their effect on the growth of these cells through DNA damage.

Yeast

Systems similar to Ames test have been developed in yeast. Saccharomyces cerevisiae is general used. These systems can check for forward and reverse mutations, as well as recombinant events.

Drosophila

Sex-Linked Recessive Lethal Test - Males from a strain with yellow bodies are used in this test. The gene for the yellow body lies on the X-chromosome. The fruit flies are fed on a diet of test chemical, and progenies are separated by sex. The surviving males are crossed with the females of the same generation, and if no males with yellow bodies are detected in the second generation, it would indicate a lethal mutation on the X-chromosome has occurred.

Plant Assays

Plants such as Zea maize, Arabidopsis and Tradescantia have been used in various test assays for mutagenecity of chemicals.

Cell Culture Assay

Mammalian cell lines such as Chinese hamster V79 cells, Chinese hampter ovary (CHO) cells or mouse lymphoma cells may be used to test for mutagenesis. Such systems include the HPRT assay for resistance to 8-azaguanine or 6-thioguanine, and ouabain-resistance (OUA) assay.

Rat primary hepatocytes may also be used to measure DNA repair following DNA damage. Mutagens may stimulate unscheduled DNA synthesis which results in more stained nuclear material in cells following exposure to mutagens.

Chromosome Check Systems

These systems check for large scale changes to the chromosomes and may be used with cell culture or in animal test. The chromosomes are stained and observed for any changes. Sister chromatid exchange is a symmetrical exchange of chromosome material between sister chromatids and may be correlated to the mutagenic or carcinogenic potential of a chemical. In micronucleus Test, cells are examined for micronuclei which are fragments or chromosomes left behind at anaphase, and is therefore a test for clastogenic agents that cause chromosome breakages. Other test may check for various chromosomal aberrations such as chromatid and chromosomal gaps and deletions, translocations, and ploidy.

Animal Test Systems

Rodents are usually used for test. The chemicals under test are usually administered in the food and in the drinking water, but sometimes by dermal application, by gavage, or by inhalation, and carried out over the major part of the life span for rodents. In tests that check for carcinogens, maximum tolerated dosage is first determined, then a range of doses are given to around 50 animals throughout the notional lifespan of the animal of two years. After death the animals are examined for sign of tumours.

Differences in metabolism between rat and human however means that human may not respond in exactly the same way to mutagen, and dosages that produce tumors on the animal test may also be unreasonably high for a human, i.e. the equivalent amount required to produce tumors in human may far exceed what a person might encounter in real life.

Mice with recessive mutations for a visible phenotype may also be used to check for mutagens. Females with recessive mutation crossed with wild-type males would yield the same phenotype as the wild-type, and any observable change to the phenotype would indicate that a mutation induced by the mutagen has occurred.

Mice may also be used for dominant lethal assays where early embryonic deaths are monitored. Male mice are treated with chemicals under test, mated with females, and the females are then sacrificed before parturition and early fetal deaths are counted in the uterine horns.

Transgenic Mouse Assay using a mouse strain infected with a viral shuttle vector is another method for testing mutagens. Animals are first treated with suspected mutagen, the mouse DNA is then isolated and the phage segment recovered and used to infect E. coli. Using similar method as the blue-white screen, the plaque formed with DNA containing mutation are white, while those without are blue.

Use of Mutagen in Anti-cancer Therapy

Many mutagens are highly toxic to proliferating cells, and they are often used to destroy cancer cells. Alkylating agents such as cyclophosphamide and cisplatin, as well as intercalating agent such asdaunorubicin and doxorubicin may be used in chemotherapy. Ionizing radiations are used in radiation therapy.

Mutagens in Fiction

In science fiction, mutagens are often represented as substances that are capable of completely changing the form of the recipient. This is seen in the Teenage Mutant Ninja Turtle franchise, comic books such as Marvel Comics's Inhumans, television series, computer and video games, like the The Witcher, Metroid Prime Trilogy, Resistance: Fall of Man, Resident Evil, Infamous, and Command & Conquer, and even toys.

Vaccine

A vaccine is a biological preparation that improves immunity to a particular disease. A vaccine typically contains an agent that resembles a disease-causing microorganism, and is often made from weakened or killed forms of the microbe or its toxins. The agent stimulates the body's immune system to recognize the agent as foreign, destroy it, and "remember" it, so that the immune system can more easily recognize and destroy any of these microorganisms that it later encounters.

Vaccines can be prophylactic (example: to prevent or ameliorate the effects of a future infection by any natural or "wild" pathogen), or therapeutic.

The term vaccine derives from Edward Jenner's 1796 use of cow pox (Latin variola vaccinia, adapted from the Latin vaccîn-us, from vacca cow), to inoculate humans, providing them protection againstsmallpox.

History

Prior to vaccination, inoculation was practised, and brought to the West in 1721 by Lady Mary Wortley Montagu, who showed it to Hans Sloane, the King's physician.

Sometime during the 1770s Edward Jenner heard a milkmaid boast that she would never have the often-fatal or disfiguring disease smallpox, because she had already had cowpox, which has a very mild effect in humans. In 1796, Jenner took pus from the hand of a milkmaid with cowpox, inoculated an 8-year-old boy with it, and six weeks later variolated the boy's arm with smallpox, afterwards observing that the boy did not catch smallpox. Further experimentation demonstrated the efficacy of the procedure on an infant. Since vaccination with cowpox was much safer than smallpox inoculation, the latter, though still widely practiced in England, was banned in 1840. Louis Pasteur generalized Jenner's idea by developing what he called a rabies vaccine, and in the nineteenth century vaccines were considered a matter of national prestige, and compulsory vaccination laws were passed.

The twentieth century saw the introduction of several successful vaccines, including those against diphtheria, measles, mumps, and rubella. Major achievements included the development of the polio vaccine in the 1950s and the eradication of smallpox during the 1960s and 1970s. Maurice Hilleman was the most prolific of the developers of the vaccines in the twentieth century. As vaccines became more common, many people began taking them for granted. However, vaccines remain elusive for many important diseases, including malaria and HIV.

Types

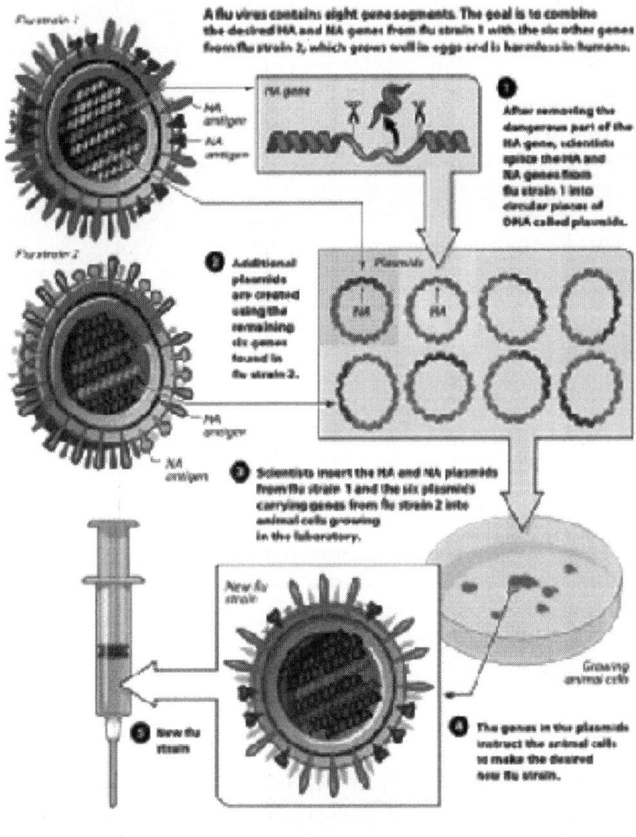

Fig. 1: Avian flu vaccine development by reverse geneticstechniques.

Vaccines are dead or inactivated organisms or purified products derived from them.

There are several types of vaccines in use. These represent different strategies used to try to reduce risk of illness, while retaining the ability to induce a beneficial immune response.

Killed

Some vaccines contain killed, but previously virulent, micro-organisms that have been destroyed with chemicals, heat, radioactivity or antibiotics. Examples are the influenza vaccine, cholera vaccine, bubonic plague vaccine, polio vaccine, hepatitis A vaccine, and rabies vaccine.

Attenuated

Some vaccines contain live, attenuated microorganisms. Many of these are live viruses that have been cultivated under conditions that disable their virulent properties, or which use closely related but less dangerous organisms to produce a broad immune response; however, some are bacterial in nature. They typically provoke more durable immunological responses and are the preferred type for healthy adults. Examples include the viral diseases yellow fever, measles, rubella, and mumps and the bacterial disease typhoid. The live Mycobacterium tuberculosisvaccine developed by Calmette and Guérin is not made of a contagious strain, but contains a virulently modified strain called "BCG" used to elicit an immune response to the vaccine. The live attenuated vaccine containing strain Yersinia pestis EV is used for plague immunization.

Toxoid

Toxoid vaccines are made from inactivated toxic compounds that cause illness rather than the micro-organism. Examples of toxoid-based vaccines include tetanus and diphtheria. Toxoid vaccines are known for their efficacy. Not all toxoids are for micro-organisms; for example,Crotalus atrox toxoid is used to vaccinate dogs against rattlesnake bites.

Subunit

Protein subunit – rather than introducing an inactivated or attenuated micro-organism to an immune system (which would constitute a "whole-agent" vaccine), a fragment of it can create an immune response. Examples include the subunit vaccine against Hepatitis B virus that is composed of only the surface proteins of the virus (previously extracted from the blood serum of chronically infected patients, but now produced by recombination of the viral genes into yeast), the virus-like particle (VLP) vaccine against human papillomavirus (HPV) that is composed of the viral major capsid protein, and the hemagglutinin and neuraminidase subunits of the influenza virus. Subunit vaccine is being used for plague immunization.

Conjugate

Conjugate – certain bacteria have polysaccharide outer coats that are poorly immunogenic. By linking these outer coats to proteins (e.g. toxins), the immune system can be led to recognize the polysaccharide as if it were a protein antigen. This approach is used in the Haemophilus influenzae type B vaccine.

Experimental

A number of innovative vaccines are also in development and in use:

§ Dendritic cell vaccines combine dendritic cells with antigens in order to present the antigens to the body's white blood cells, thus stimulating an immune reaction. These vaccines have shown some positive preliminary results for treating brain tumors.

§ Recombinant Vector – by combining the physiology of one micro-organism and the DNA of the other, immunity can be created against diseases that have complex infection processes

§ DNA vaccination – in recent years a new type of vaccine called DNA vaccination, created from an infectious agent's DNA, has been developed. It works by insertion (and expression, triggering immune system recognition) of viral or bacterial DNA into human or animal cells. Some cells of the immune system that recognize the proteins expressed will mount an attack against these proteins and cells expressing them. Because these cells live for a very long time, if the pathogen that normally expresses these proteins is encountered at a later time, they will be attacked instantly by the immune system. One advantage of DNA vaccines is that they are very easy to produce and store. As of 2006, DNA vaccination is still experimental.

§ T-cell receptor peptide vaccines are under development for several diseases using models of Valley Fever, stomatitis, and atopic dermatitis. These peptides have been shown to modulate cytokineproduction and improve cell mediated immunity.

§ Targeting of identified bacterial proteins that are involved in complement inhibition would neutralize the key bacterial virulence mechanism.

While most vaccines are created using inactivated or attenuated compounds from micro-organisms, synthetic vaccines are composed mainly or wholly of synthetic peptides, carbohydrates or antigens.

Valence

Vaccines may be monovalent (also called univalent) or multivalent (also called polyvalent). A monovalent vaccine is designed to immunize against a single antigen or single microorganism. A multivalent or polyvalent vaccine is designed to immunize against two or more strains of the same microorganism, or against two or more microorganisms. In certain cases a monovalent vaccine may be preferable for rapidly developing a strong immune response.

Developing Immunity

The immune system recognizes vaccine agents as foreign, destroys them, and "remembers" them. When the virulent version of an agent comes along the body recognizes the protein coat on the virus, and thus is prepared to respond, by (1) neutralizing the target agent before it can enter cells, and (2) by recognizing and destroying infected cells before that agent can multiply to vast numbers.

When two or more vaccines are mixed together in the same formulation, the two vaccines can interfere. This most frequently occurs with live attenuated vaccines, where one of the vaccine components is more robust than the others and suppresses the growth and immune response to the other components. This phenomenon was first noted in the trivalent Sabin polio vaccine, where the amount of serotype 2 virus in the vaccine had to be reduced to stop it from interfering with the "take" of the serotype 1 and 3 viruses in the vaccine. This phenomenon has also been found to be a problem with the dengue vaccines currently being researched, where the DEN-3 serotype was found to predominate and suppress the response to DEN-1, -2 and -4 serotypes.

Vaccines have contributed to the eradication of smallpox, one of the most contagious and deadly diseases known to man. Other diseases such as rubella, polio, measles, mumps, chickenpox, andtyphoid are nowhere near as common as they were a hundred years ago. As long as the vast majority of people are vaccinated, it is much more difficult for an outbreak of disease to occur, let alone spread. This effect is called herd immunity. Polio, which is transmitted only between humans, is targeted by an extensive eradication campaign that has seen endemic polio restricted to only parts of four countries (Afghanistan, India, Nigeria and Pakistan).[2] The difficulty of reaching all children as well as cultural misunderstandings, however, have caused the anticipated eradication date to be missed several times.

Schedule

In order to provide best protection, children are recommended to receive vaccinations as soon as their immune systems are sufficiently developed to respond to particular vaccines, with additional "booster" shots often required to achieve "full immunity". This has led to the development of complex vaccination schedules.

In the United States, the Advisory Committee on Immunization Practices, which recommends schedule additions for the Centers for Disease Control and Prevention, recommends routine vaccination of children against: hepatitis A, hepatitis B, polio, mumps, measles, rubella,diphtheria, pertussis, tetanus, HiB, chickenpox, rotavirus, influenza, meningococcal disease and pneumonia. The large number of vaccines and boosters recommended (up to 24 injections by age two) has led to problems with achieving full compliance. In order to combat declining compliance rates, various notification systems have been instituted and a number of combination injections are now marketed (e.g., Pneumococcal conjugate vaccine and MMRV vaccine), which provide protection against multiple diseases.

Besides recommendations for infant vaccinations and boosters, many specific vaccines are recommended at other ages or for repeated injections throughout life—most commonly for measles, tetanus, influenza, and pneumonia. Pregnant women are often screened for continued resistance to rubella. The human papillomavirus vaccine is recommended in the U.S. (as of 2011) and UK (as of 2009). Vaccine recommendations for the elderly concentrate on pneumonia and influenza, which are more deadly to that group. In 2006, a vaccine was introduced against shingles, a disease caused by the chickenpox virus, which usually affects the elderly.

Effectiveness

Vaccines do not guarantee complete protection from a disease. Sometimes, this is because the host's immune system simply does not respond adequately or at all. This may be due to a lowered immunity in general (diabetes, steroid use, HIV infection, age) or because the host's immune system does not have a B cell capable of generating antibodies to that antigen.

Even if the host develops antibodies, the human immune system is not perfect and in any case the immune system might still not be able to defeat the infection immediately. In this case, the infection will be less severe and heal faster.

Adjuvants are typically used to boost immune response. Most often aluminium adjuvants are used, but adjuvants like squalene are also used in some vaccines and more vaccines with squalene and phosphate adjuvants are being tested. Larger doses are used in some cases for older people (50–75 years and up), whose immune response to a given vaccine is not as strong.

The efficacy or performance of the vaccine is dependent on a number of factors:

- § the disease itself (for some diseases vaccination performs better than for other diseases)
- § the strain of vaccine (some vaccinations are for different strains of the disease)
- § whether one kept to the timetable for the vaccinations.
- § some individuals are "non-responders" to certain vaccines, meaning that they do not generate antibodies even after being vaccinated correctly
- § other factors such as ethnicity, age, or genetic predisposition.

When a vaccinated individual does develop the disease vaccinated against, the disease is likely to be milder than without vaccination.

The following are important considerations in the effectiveness of a vaccination program:

1. careful modelling to anticipate the impact that an immunization campaign will have on the epidemiology of the disease in the medium to long term

2. ongoing surveillance for the relevant disease following introduction of a new vaccine and
3. maintaining high immunization rates, even when a disease has become rare.

In 1958 there were 763,094 cases of measles and 552 deaths in the United States. With the help of new vaccines, the number of cases dropped to fewer than 150 per year (median of 56). In early 2008, there were 64 suspected cases of measles. 54 out of 64 infections were associated with importation from another country, although only 13% were actually acquired outside of the United States; 63 of these 64 individuals either had never been vaccinated against measles, or were uncertain whether they had been vaccinated.

Opposition to Vaccination

Fig. 2: James Gillray, The Cow-Pock—or—the Wonderful Effects of the New Inoculation!(1802)

Opposition to vaccination, from a wide array of vaccine critics, has existed since the earliest vaccination campaigns. Although the benefits of preventing suffering and death from serious infectious diseases greatly outweigh the risks of rare adverse effects following immunization, disputes have arisen over the morality, ethics, effectiveness, and safety of vaccination. Some vaccination critics say that vaccines are ineffective against disease or that vaccine safety studies are inadequate. Some religious groups do not allow vaccination, and some political groups oppose mandatory vaccination on the grounds of individual liberty. In response, concern has been raised that spreading unfounded information about the medical risks of vaccines increases rates of life-threatening infections, not only in the children whose parents refused vaccinations, but also in other children, perhaps too young for vaccines, who could contract infections from unvaccinated carriers.

Economics of Development

One challenge in vaccine development is economic: many of the diseases most demanding a vaccine, including HIV, malaria and tuberculosis, exist principally in poor countries. Pharmaceutical firms and biotechnology companies have little incentive to develop vaccines for these diseases, because there is little revenue potential. Even in more affluent countries, financial returns are usually minimal and the financial and other risks are great.

Most vaccine development to date has relied on "push" funding by government, universities and non-profit organizations. Many vaccines have been highly cost effective and beneficial for public health. The number of vaccines actually administered has risen dramatically in recent decades. This increase, particularly in the number of different vaccines administered to children before entry into schools may be due to government mandates and support, rather than economic incentive.

Patents

The filing of patents on vaccine development processes can also be viewed as an obstacle to the development of new vaccines. Because of the weak protection offered through a patent on the final product, the protection of the innovation regarding vaccines is often made through the patent of processes used on the development of new vaccines as well as the protection of secrecy.

Production

Vaccine production has several stages. First, the antigen itself is generated. Viruses are grown either on primary cells such as chicken eggs (e.g., for influenza), or on continuous cell lines such as cultured human cells (e.g., for hepatitis A). Bacteria are grown in bioreactors (e.g., Haemophilus influenzae type b). Alternatively, a recombinant protein derived from the viruses or bacteria can be generated in yeast, bacteria, or cell cultures. After the antigen is generated, it is isolated from the cells used to generate it. A virus may need to be inactivated, possibly with no further purification required. Recombinant proteins need many operations involving ultrafiltration and column chromatography. Finally, the vaccine is formulated by adding adjuvant, stabilizers, and preservatives as needed. The adjuvant enhances the immune response of the antigen, stabilizers increase the storage life, and preservatives allow the use of multidose vials. Combination vaccines are harder to develop and produce, because of potential incompatibilities and interactions among the antigens and other ingredients involved.

Vaccine production techniques are evolving. Cultured mammalian cells are expected to become increasingly important, compared to conventional options such as chicken eggs, due to greater productivitity and low incidence of problems with contamination.

Recombination technology that produces genetically detoxified vaccine is expected to grow in popularity for the production of bacterial vaccines that use toxoids. Combination vaccines are expected to reduce the quantities of antigens they contain, and thereby decrease undesirable interactions, by using pathogen-associated molecular patterns.

Excipients

Beside the active vaccine itself, the following excipients are commonly present in vaccine preparations:

§ Aluminum salts or gels are added as adjuvants. Adjuvants are added to promote an earlier, more potent response, and more persistent immune response to the vaccine; they allow for a lower vaccine dosage.

§ Antibiotics are added to some vaccines to prevent the growth of bacteria during production and storage of the vaccine.

§ Egg protein is present in influenza and yellow fever vaccines as they are prepared using chicken eggs. Other proteins may be present.

§ Formaldehyde is used to inactivate bacterial products for toxoid vaccines. Formaldehyde is also used to kill unwanted viruses and bacteria that might contaminate the vaccine during production.

§ Monosodium glutamate (MSG) and 2-phenoxyethanol are used as stabilizers in a few vaccines to help the vaccine remain unchanged when the vaccine is exposed to heat, light, acidity, or humidity.

§ Thimerosal is a mercury-containing preservative that is added to vials of vaccine that contain more than one dose to prevent contamination and growth of potentially harmful bacteria.

Role of Preservatives

Many vaccines need preservatives to prevent serious adverse effects such as Staphylococcus infection that, in one 1928 incident, killed 12 of 21 children inoculated with a diphtheria vaccine that lacked a preservative. Several preservatives are available, including thiomersal, phenoxyethanol, and formaldehyde. Thiomersal is more effective against bacteria, has better shelf life, and improves vaccine stability, potency, and safety, but in the U.S., the European Union, and a few other affluent countries, it is no longer used as a preservative in childhood vaccines, as a precautionary measure due to itsmercury content. Although controversial claims have been made that thiomersal contributes to autism, no convincing scientific evidence supports these claims.

Delivery Systems

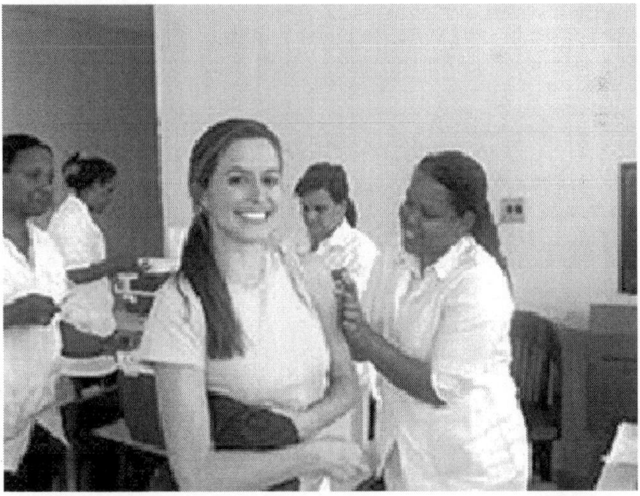

Fig. 3: Woman receiving rubella vaccination, Brazil, 2008.

There are several new delivery systems in development that will hopefully make vaccines more efficient to deliver. Possible methods include liposomesand ISCOM (immune stimulating complex).

The latest developments in vaccine delivery technologies have resulted in oral vaccines. A polio vaccine was developed and tested by volunteer vaccinations with no formal training; the results were very positive in that the ease of the vaccines increased dramatically. With an oral vaccine, there is no risk of blood contamination. Oral vaccines are likely to be solid which have proven to be more stable and less likely to freeze; this stability reduces the need for a "cold chain": the resources required to keep vaccines within a restricted temperature range from the manufacturing stage to the point of administration, which, in turn, will decrease costs of vaccines. Finally, a microneedle approach, which is still in stages of development, seems to be the vaccine of the future, the microneedle, which is "pointed projections fabricated into arrays that can create vaccine delivery pathways through the skin".

Plasmids

The use of plasmids has been validated in preclinical studies as a protective vaccine strategy for cancer and infectious diseases. However, in human studies this approach has failed to provide clinically relevant benefit. The overall efficacy of plasmid DNA immunization depends on increasing the plasmid's immunogenicity while also correcting for factors involved in the specific activation of immune effector cells.

Use in Veterinary Medicine

Vaccinations of animals are used both to prevent their contracting diseases and to prevent transmission of disease to humans. Both animals kept as pets and animals raised as livestock are routinely vaccinated. In some instances, wild populations may be vaccinated. This is sometimes accomplished with vaccine-laced food spread in a disease-prone area and has been used to attempt to control rabies in raccoons.

Where rabies occurs, rabies vaccination of dogs may be required by law. Other canine vaccines include canine distemper, canine parvovirus, infectious canine hepatitis, adenovirus-2, leptospirosis,bordatella, canine parainfluenza virus, and Lyme disease among others.

Trends

Vaccine development has several trends:

§ Until recently, most vaccines were aimed at infants and children, but adolescents and adults are increasingly being targeted.

§ Combinations of vaccines are becoming more common; vaccines containing five or more components are used in many parts of the world.

§ New methods of administering vaccines are being developed, such as skin patches, aerosols via inhalation devices, and eating genetically engineered plants.

§ Vaccines are being designed to stimulate innate immune responses, as well as adaptive.

§ Attempts are being made to develop vaccines to help cure chronic infections, as opposed to preventing disease.

§ Vaccines are being developed to defend against bioterrorist attacks such as anthrax, plague, and smallpox.

§ Appreciation for sex and pregnancy differences in vaccine responses "might change the strategies used by public health officials".

Principles that govern the immune response can now be used in tailor-made vaccines against many noninfectious human diseases, such as cancers and autoimmune disorders. For example, the experimental vaccine CYT006-AngQb has been investigated as a possible treatment for high blood pressure. Factors that have impact on the trends of vaccine development include progress in translatory medicine, demographics, regulatory science, political, cultural, and social responses.

Chapter 6

Genetic Engineering

Genetic engineering, also called genetic modification, is the direct human manipulation of an organism's genome using modern DNA technology. It involves the introduction offoreign DNA or synthetic genes into the organism of interest. The introduction of new DNA does not require the use of classical genetic methods, however traditional breedingmethods are typically used for the propagation of recombinant organisms.

An organism that is generated through the introduction of recombinant DNA is considered to be a genetically modified organism. The first organisms genetically engineered were bacteria in 1973 and then mice in 1974. Insulin-producing bacteria were commercialized in 1982 and genetically modified food has been sold since 1994.

The most common form of genetic engineering involves the insertion of new genetic material at an unspecified location in the host genome. This is accomplished by isolating and copying the genetic material of interest using molecular cloning methods to generate a DNA sequence containing the required genetic elements for expression, and then inserting this construct into the host organism. Other forms of genetic engineering include gene targeting and knocking out specific genes via engineered nucleases such aszinc finger nucleases or engineered homing endonucleases.

Genetic engineering techniques have been applied in numerous fields including research, biotechnology, and medicine. Medicines such as insulin and human growth hormone are now produced in bacteria, experimental mice such as the oncomouse and the knockout mouse are being used for research purposes and insect resistant and/or herbicide tolerant crops have been commercialized. Genetically engineered plants and animals capable of producing biotechnology drugs more cheaply than current methods (calledpharming) are also being developed and in 2009 the FDA approved the sale of the pharmaceutical protein antithrombin produced in the milk of genetically engineered goats.

Definition

Genetic engineering alters the genetic makeup of an organism using techniques that introduce heritable material prepared outside the organism either directly into the host or into a cell that is thenfused or hybridized with the host. This involves using recombinant nucleic acid (DNA or RNA) techniques to form new combinations of heritable genetic material followed by the incorporation of that material either indirectly through a vector system or directly through micro-injection, macro-injection and micro-encapsulation techniques. Genetic engineering does not include traditional animal andplant breeding, in vitro fertilisation, induction of polyploidy, mutagenesis and cell fusion techniques that do not use recombinant nucleic acids or a genetically modified organism in the process. Cloning and stem cell research, although not considered genetic engineering, are closely related and genetic engineering can be used within them. Synthetic biology is an emerging discipline that takes genetic engineering a step further by introducing artificially synthesized genetic material from raw materials into an organism.

If genetic material from another species is added to the host, the resulting organism is called transgenic. If genetic material from the same species or a species that can naturally breed with the host is used the resulting organism is called cisgenic. Genetic engineering can also be used to remove genetic material from the target organism, creating a knock out organism. In Europe genetic modification is synonymous with genetic engineering while within the United States of America it can also refer to conventional breeding methods. Within the scientific community, the term genetic engineering is not commonly used; more specific terms such as transgenic are preferred.

History

Humans have altered the genomes of species for thousands of years through artificial selection and more recently mutagenesis. Genetic engineering as the direct manipulation of DNA by humans outside breeding and mutations has only existed since the 1970s. The term "genetic engineering" was first coined by Jack Williamson in his science fiction novel Dragon's Island, published in 1951,one year before DNA's role in heredity was confirmed by Alfred Hershey and Martha Chase, and two years before James Watson and Francis Crick showed that the DNA molecule has a double-helix structure.

In 1972 Paul Berg created the first recombinant DNA molecules by combining DNA from the monkey virus SV40 with that of the lambda virus. In 1973 Herbert Boyer and Stanley Cohen created the first transgenic organism by inserting antibiotic resistance genes into the plasmid of an E. coli bacterium. A year later Rudolf Jaenisch created a transgenic mouse by introducing foreign DNA into its embryo, making it the world's first transgenic animal.

In 1976 Genentech, the first genetic engineering company was founded by Herbert Boyer and Robert Swanson and a year later and the company produced a human protein (somatostatin) in E.coli. Genentech announced the production of genetically engineered human insulin in 1978. In 1980, the U.S. Supreme Court in the Diamond v. Chakrabarty case ruled that genetically altered life could be patented. The insulin produced by bacteria, branded humulin, was approved for release by the Food and Drug Administration in 1982.

The first field trials of genetically engineered plants occurred in France and the USA in 1986, tobacco plants were engineered to be resistant to herbicides. The People's Republic of China was the first country to commercialize transgenic plants, introducing a virus-resistant tobacco in 1992.

In 1994 Calgene attained approval to commercially release the Flavr Savr tomato, a tomato engineered to have a longer shelf life. In 1994, the European Union approved tobacco engineered to be resistant to the herbicide bromoxynil, making it the first genetically engineered crop commercialized in Europe.

In 1995, Bt Potato was approved safe by the Environmental Protection Agency, making it the first pesticide producing crop to be approved in the USA. In 2009 11 transgenic crops were grown commercially in 25 countries, the largest of which by area grown were the USA, Brazil, Argentina, India, Canada, China, Paraguay and South Africa.

In 2010, scientists at the J. Craig Venter Institute, announced that they had created the first synthetic bacterial genome, and added it to a cell containing no DNA. The resulting bacterium, namedSynthia, was the world's first synthetic life form.

Process

Isolating the Gene

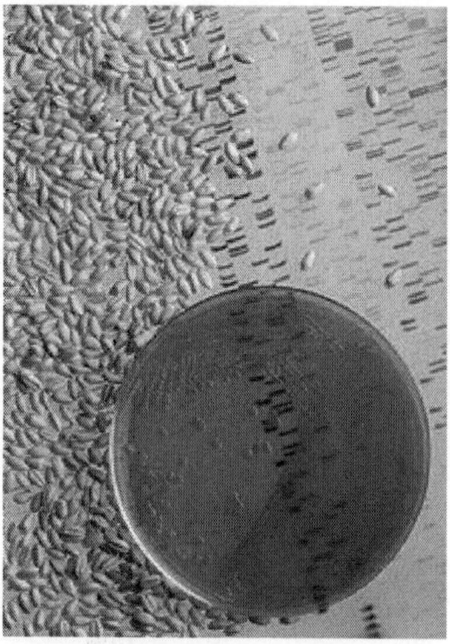

Fig. 1: Elements of genetic engineering

First, the gene to be inserted into the genetically modified organism must be chosen and isolated. Presently, most genes transferred into plants provide protection against insects or tolerance to herbicides. In animals the majority of genes used are growth hormone genes. Once chosen the genes must be isolated. This typically involves multiplying the gene using polymerase chain reaction (PCR). If the chosen gene or the donor organism's genome has been well studied it may be present in a genetic library. If the DNA sequence is known, but no copies of the gene are available, it can be artificially synthesized. Once isolated, the gene is inserted into a bacterial plasmid.

Constructs

The gene to be inserted into the genetically modified organism must be combined with other genetic elements in order for it to work properly. The gene can also be modified at this stage for better expression or effectiveness. As well as the gene to be inserted most constructs contain a promoter and terminatorregion as well as a selectable marker gene. The promoter region initiates transcription of the gene and can be used to control the location and level of gene expression, while the terminator region ends transcription. The selectable marker, which in most cases confers antibiotic resistance to the organism it is expressed in, is needed to determine which cells are transformed with the new gene. The constructs are made using recombinant DNA techniques, such asrestriction digests, ligations and molecular cloning.

Gene Targeting

The most common form of genetic engineering involves inserting new genetic material randomly within the host genome. Other techniques allow new genetic material to be inserted at a specific location in the host genome or generate mutations at desired genomic loci capable of knocking out endogenous genes. The technique of gene targeting uses homologous recombination to target desired changes to a specific endogenous gene. This tends to occur at a relatively low frequency in plants and animals and generally requires the use of selectable markers. The frequency of gene targeting can be greatly enhanced with the use of engineered nucleases such as zinc finger nucleases, engineered homing endonucleases, or nucleases created from TAL effectors. In addition to enhancing gene targeting, engineered nucleases can also be used to introduce mutations at endogenous genes that generate a gene knockout .

Transformation

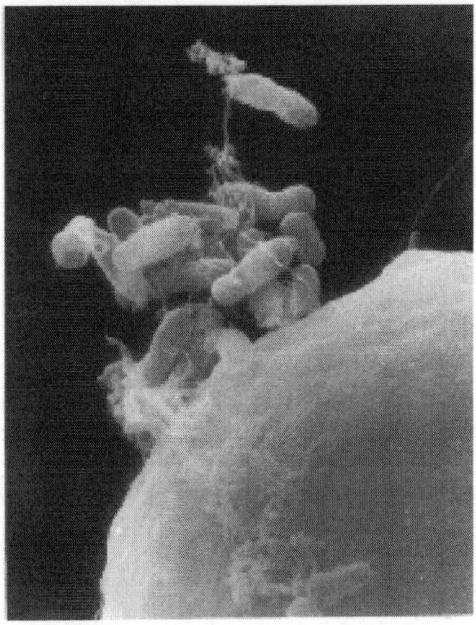

Fig. 2: A. tumefaciens attaching itself to a carrot cell

About 1% of bacteria are naturally able to take up foreign DNA but it can also be induced in other bacteria. Stressing the bacteria for example, with a heat shockor an electric shock, can make the cell membrane permeable to DNA that may then incorporate into their genome or exist as extrachromosomal DNA. DNA is generally inserted into animal cells using microinjection, where it can be injected through the cells nuclear envelope directly into the nucleus or through the use ofviral vectors. In plants the DNA is generally inserted using Agrobacterium-mediated recombination or biolistics.

In Agrobacterium-mediated recombination the plasmid construct must also contain T-DNA. Agrobacterium naturally inserts DNA from a tumor inducing plasmid into any susceptible plant's genome it infects, causing crown gall disease. The T-DNA region of this plasmid is responsible for insertion of the DNA. The genes to be inserted are cloned into a binary vector, which contains T-DNA and can be grown in both E. Coli and Agrobacterium. Once the binary vector is constructed the plasmid is transformed into Agrobacterium containing no plasmids and plant cells are infected. The Agrobacterium will then naturally insert the genetic material into the plant cells.

In biolistics particles of gold or tungsten are coated with DNA and then shot into young plant cells or plant embryos. Some genetic material will enter the cells and transform them. This method can be used on plants that are not susceptible to Agrobacterium infection and also allows transformation of plant plastids. Another transformation method for plant and animal cells is electroporation. Electroporation involves subjecting the plant or animal cell to an electric shock, which can make the cell membrane permeable to plasmid DNA. In some cases the electroporated cells will incorporate the DNA into their genome. Due to the damage caused to the cells and DNA the transformation efficiency of biolistics and electroporation is lower than agrobacterial mediated transformation and microinjection.

Selection

Not all the organism's cells will be transformed with the new genetic material; in most cases a selectable marker is used to differentiate transformed from untransformed cells. If a cell has been successfully transformed with the DNA it will also contain the marker gene. By growing the cells in the presence of an antibiotic or chemical that selects or marks the cells expressing that gene it is possible to separate the transgenic events from the non-transgenic. Another method of screening involves using a DNA probe that will only stick to the inserted gene. A number of strategies have been developed that can remove the selectable marker from the mature transgenic plant.

Regeneration

As often only a single cell is transformed with genetic material the organism must be regrown from that single cell. As bacteria consist of a single cell and reproduce clonally regeneration is not necessary. In plants this is accomplished through the use of tissue culture. Each plant species has different requirements for successful regeneration through tissue culture. If successful an adult plant is produced that contains the transgene in every cell. In animals it is necessary to ensure that the inserted DNA is present in the embryonic stem cells. When the offspring is produced they can be screened for the presence of the gene. All offspring from the first generation will be heterozygous for the inserted gene and must be mated together to produce a homozygous animal.

Confirmation

The finding that a recombinant organism contains the inserted genes is not usually sufficient to ensure that the genes will be expressed in an appropriate manner in the intended tissues of the recombinant organism. To examine the presence of the gene, further analysis frequently uses PCR, Southern hybridization, and DNA sequencing, which serve to determine the chromosomal location and copy number of the inserted gene. To examine expression of the trans-gene, an extensive analysis of transcription, RNA processing patterns, and the expression and localization of the protein product(s) is usually necessary, using methods including northern hybridization, quantitative RT-PCR, Western blot, immunofluorescence and phenotypic analysis. When appropriate, the organism's offspring are studied to confirm that the trans-gene and associated phenotype are stably inherited.

Applications

Genetic engineering has applications in medicine, research, industry and agriculture and can be used on a wide range of plants, animals and micro organism.

Medicine

In medicine genetic engineering has been used to mass-produce insulin, human growth hormones, follistim (for treating infertility), human albumin, monoclonal antibodies, antihemophilic factors,vaccines and many other drugs. Vaccination generally involves injecting weak live, killed or inactivated forms of viruses or their toxins into the person being immunized. Genetically engineered viruses are being developed that can still confer immunity, but lack the infectious sequences. Mouse hybridomas, cells fused together to create monoclonal antibodies, have been humanised through genetic engineering to create human monoclonal antibodies.

Genetic engineering is used to create animal models of human diseases. Genetically modified mice are the most common genetically engineered animal model. They have been used to study and model cancer (the oncomouse), obesity, heart disease, diabetes, arthritis, substance abuse, anxiety, aging and Parkinson disease. Potential cures can be tested against these mouse models. Also genetically modified pigs have been bred with the aim of increasing the success of pig to human organ transplantation.

Gene therapy is the genetic engineering of humans by replacing defective human genes with functional copies. This can occur in somatic tissue or germline tissue. If the gene is inserted into the germline tissue it can be passed down to that person's descendants. Gene therapy has been used to treat patients suffering from immune deficiencies (notably Severe combined immunodeficiency) and trials have been carried out on other genetic disorders. The success of gene therapy so far has been limited and a patient (Jesse Gelsinger) has died during a clinical trial testing a new treatment. There are also ethical concerns should the technology be used not just for treatment, but for enhancement, modification or alteration of a human beings' appearance, adaptability, intelligence, character or behavior. The distinction between cure and enhancement can also be difficult to establish. Transhumanists consider the enhancement of humans desirable.

Research

Fig. 3 : Knockout mice

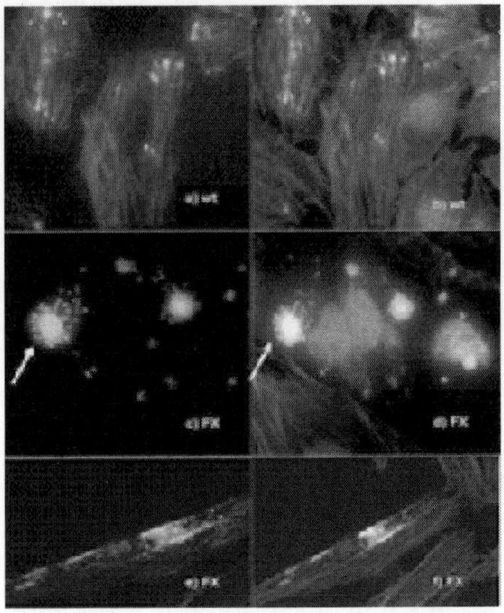

Fig. 4: Human cells in which some proteins are fused with green fluorescent protein to allow them to be visualised

Genetic engineering is an important tool for natural scientists. Genes and other genetic information from a wide range of organisms are transformed into bacteria for storage and modification, creating genetically modified bacteria in the process. Bacteria are cheap, easy to grow, clonal, multiply quickly, relatively easy to transform and can be stored at -80 °C almost indefinitely. Once a gene is isolated it can be stored inside the bacteria providing an unlimited supply for research.

Organisms are genetically engineered to discover the functions of certain genes. This could be the effect on the phenotype of the organism, where the gene is expressed or what other genes it interacts with. These experiments generally involve loss of function, gain of function, tracking and expression.

§ Loss of function experiments, such as in a gene knockout experiment, in which an organism is engineered to lack the activity of one or more genes. A knockout experiment involves the creation and manipulation of a DNA construct in vitro, which, in a simple knockout, consists of a copy of the desired gene, which has been altered such that it is non-functional. Embryonic stem cells incorporate the altered gene, which replaces the already present functional copy. These stem cells are injected into blastocysts, which are implanted into surrogate mothers. This allows the experimenter to analyze the defects caused by this mutation and thereby determine the role of particular genes. It is used especially frequently in developmental biology. Another method, useful in organisms such as Drosophila (fruit fly), is to induce mutations in a large population and then screen the progeny for the desired mutation. A similar process can be used in both plants and prokaryotes.

§ Gain of function experiments, the logical counterpart of knockouts. These are sometimes performed in conjunction with knockout experiments to more finely establish the function of the desired gene. The process is much the same as that in knockout engineering, except that the construct is designed to increase the function of the gene, usually by providing extra copies of the gene or inducing synthesis of the protein more frequently.

§ Tracking experiments, which seek to gain information about the localization and interaction of the desired protein. One way to do this is to replace the wild-type gene with a 'fusion' gene, which is a juxtaposition of the wild-type gene with a reporting element such as green fluorescent protein (GFP) that will allow easy visualization of the products of the genetic modification. While this is a useful technique, the manipulation can destroy the function of the gene, creating secondary effects and possibly calling into question the results of the experiment. More sophisticated techniques are now in development that can track protein products without mitigating their function, such as the addition of small sequences that will serve as binding motifs to monoclonal antibodies.

§ Expression studies aim to discover where and when specific proteins are produced. In these experiments, the DNA sequence before the DNA that codes for a protein, known as a gene's promoter, is reintroduced into an organism with the protein coding region replaced by a reporter gene such as GFP or an enzyme that catalyzes the production of a dye. Thus the time and place where a particular protein is produced can be observed. Expression studies can be taken a step further by altering the promoter to find which pieces are crucial for the proper expression of the gene and are actually bound by transcription factor proteins; this process is known as promoter bashing.

Industrial

By engineering genes into bacterial plasmids it is possible to create a biological factory that can produce proteins and enzymes. Some genes do not work well in bacteria, so yeast, a eukaryote, can also be used. Bacteria and yeast factories have been used to produce medicines such as insulin, human growth hormone, and vaccines, supplements such as tryptophan, aid in the production of food (chymosin in cheese making) and fuels. Other applications involving genetically engineered bacteria being investigated involve making the bacteria perform tasks outside their natural cycle, such as cleaning up oil spills, carbon and other toxic waste.

Agriculture

Fig. 5: Bt-toxins present in peanut leaves (bottom image) protect it from extensive damage caused by European corn borerlarvae (top image).

One of the best-known and controversial applications of genetic engineering is the creation of genetically modified food. There are three generations of genetically modified crops. First generation crops have been commercialized and most provide protection from insects and/or resistance to herbicides. There are also fungal and virus resistant crops developed or in development. They have been developed to make the insect and weed management of crops easier and can indirectly increase crop yield.

The second generation of genetically modified crops being developed aim to directly improve yield by improving salt, cold or drought tolerance and to increase the nutritional value of the crops.

The third generation consists of pharmaceutical crops, crops that contain edible vaccines and other drugs. Some agriculturally important animals have been genetically modified with growth hormones to increase their size while others have been engineered to express drugs and other proteins in their milk.

The genetic engineering of agricultural crops can increase the growth rates and resistance to different diseases caused by pathogens and parasites. This is beneficial as it can greatly increase the production of food sources with the usage of fewer resources that would be required to host the world's growing populations. These modified crops would also reduce the usage of chemicals, such as fertilizers and pesticides, and therefore decrease the severity and frequency of the damages produced by these chemical pollution.

Ethical and safety concerns have been raised around the use of genetically modified food. A major safety concern relates to the human health implications of eating genetically modified food, in particular whether toxic or allergic reactions could occur.

Gene flow into related non-transgenic crops, off target effects on beneficial organisms and the impact on biodiversity are important environmental issues. Ethical concerns involve religious issues, corporatecontrol of the food supply, intellectual property rights and the level of labeling needed on genetically modified products.

Other Uses

In materials science, a genetically modified virus has been used to construct a more environmentally friendly lithium-ion battery. Some bacteria have been genetically engineered to create black and white photographs while others have potential to be used as sensors by expressing a fluorescent protein under certain environmental conditions. Genetic engineering is also being used to create BioArt and novelty items such as blue roses, and glowing fish.

Opposition and Criticism

A 2010 study of Canola found transgenes in 80% of wild (uncultivated or "feral") varieties in North Dakota, meaning 80% of the plants which had established themselves in the area were genetically engineered varieties. The researchers stated that "we found the highest densities of [such transgene-containing] plants near agricultural fields and along major freeways, but we were also finding plants in the middle of nowhere" adding that "over time,..the build-up of different types of herbicide resistance in feral [natural] canola and closely related weeds, like field mustard, could make it more difficult to manage these plants using herbicides."

Chromosome

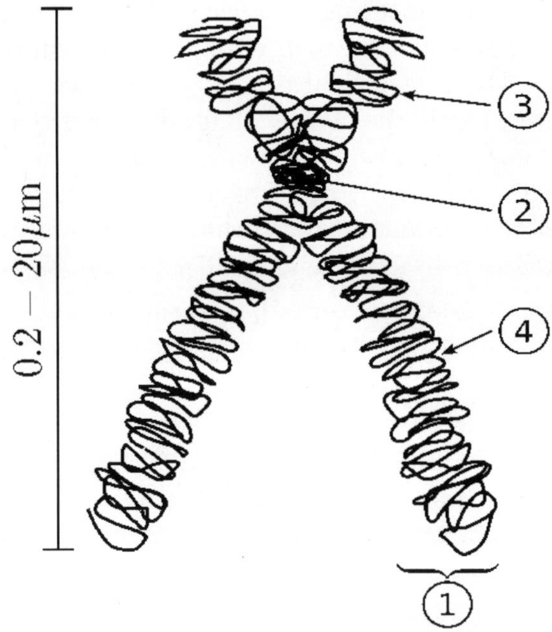

Fig. 6: Diagram of a replicated and condensedmetaphase eukaryotic chromosome. (1)Chromatid – one of the two identical parts of the chromosome after S phase. (2) Centromere – the point where the two chromatids touch, and where the microtubules attach. (3) Short arm. (4) Long arm.

A chromosome is an organized structure of DNA and protein found in cells. It is a single piece of coiled DNA containing many genes, regulatory elements and other nucleotide sequences. Chromosomes also contain DNA-bound proteins, which serve to package the DNA and control its functions.

Chromosomes vary widely between different organisms. The DNA molecule may be circular or linear, and can be composed of 100,000 to 10,000,000,000 nucleotides in a long chain. Typically, eukaryotic cells (cells with nuclei) have large linear chromosomes and prokaryotic cells (cells without defined nuclei) have smaller circular chromosomes, although there are many exceptions to this rule. Also, cells may contain more than one type of chromosome; for example, mitochondria in most eukaryotes and chloroplasts in plants have their own small chromosomes.

In eukaryotes, nuclear chromosomes are packaged by proteins into a condensed structure called chromatin. This allows the very long DNA molecules to fit into the cell nucleus. The structure of chromosomes and chromatin varies through the cell cycle. Chromosomes are the essential unit for cellular division and must be replicated, divided, and passed successfully to their daughter cells so as to ensure the genetic diversity and survival of theirprogeny. Chromosomes may exist as either duplicated or unduplicated. Unduplicated chromosomes are single linear strands, whereas duplicated chromosomes (copied during synthesis phase) contain two copies joined by a centromere.

Compaction of the duplicated chromosomes during mitosis and meiosis results in the classic four-arm structure (pictured to the right). Chromosomal recombination plays a vital role in genetic diversity. If these structures are manipulated incorrectly, through processes known as chromosomal instability and translocation, the cell may undergo mitotic catastrophe and die, or it may unexpectedly evade apoptosis leading to the progression of cancer.

In practice "chromosome" is a rather loosely defined term. In prokaryotes and viruses, the term genophore is more appropriate when no chromatin is present. However, a large body of work uses the term chromosome regardless of chromatin content. In prokaryotes, DNA is usually arranged as a circle, which is tightly coiled in on itself, sometimes accompanied by one or more smaller, circular DNA molecules called plasmids.

These small circular genomes are also found in mitochondria and chloroplasts, reflecting their bacterial origins. The simplest genophores are found in viruses: these DNA orRNA molecules are short linear or circular genophores that often lack structural proteins.

The word chromosome comes from the Greek ÷ñþìá (chroma, colour) and óþìá (soma, body) due to their property of being very strongly stained by particular dyes.

History

In a series of experiments beginning in the mid-1880s, Theodor Boveri gave the definitive demonstration that chromosomes are the vectors of heredity. His two principles were the continuity of chromosomes and the individuality of chromosomes. It is the second of these principles that was so original. Wilhelm Roux suggested that each chromosome carries a different genetic load. Boveri was able to test and confirm this hypothesis.

Aided by the rediscovery at the start of the 1900s of Gregor Mendel's earlier work, Boveri was able to point out the connection between the rules of inheritance and the behaviour of the chromosomes. Boveri influenced two generations of American cytologists: Edmund Beecher Wilson, Walter Sutton and Theophilus Painter were all influenced by Boveri (Wilson and Painter actually worked with him).

In his famous textbook The Cell in Development and Heredity, Wilson linked together the independent work of Boveri and Sutton (both around 1902) by naming the chromosome theory of inheritance the "Sutton-Boveri Theory" (the names are sometimes reversed). Ernst Mayr remarks that the theory was hotly contested by some famous geneticists: William Bateson, Wilhelm Johannsen, Richard Goldschmidt and T.H. Morgan, all of a rather dogmatic turn of mind. Eventually, complete proof came from chromosome maps in Morgan's own lab.

In Eukaryotes

Eukaryotes (cells with nuclei such as those found in plants, yeast, and animals) possess multiple large linear chromosomes contained in the cell's nucleus. Each chromosome has one centromere, with one or two arms projecting from the centromere, although, under most circumstances, these arms are not visible as such. In addition, most eukaryotes have a small circular mitochondrial genome, and some eukaryotes may have additional small circular or linear cytoplasmic chromosomes.

In the nuclear chromosomes of eukaryotes, the uncondensed DNA exists in a semi-ordered structure, where it is wrapped around histones (structural proteins), forming a composite material calledchromatin.

Chromatin

Chromatin is the complex of DNA and protein found in the eukaryotic nucleus, which packages chromosomes. The structure of chromatin varies significantly between different stages of the cell cycle, according to the requirements of the DNA.

Interphase Chromatin

During interphase (the period of the cell cycle where the cell is not dividing), two types of chromatin can be distinguished:

- § Euchromatin, which consists of DNA that is active, e.g., being expressed as protein.
- § Heterochromatin, which consists of mostly inactive DNA. It seems to serve structural purposes during the chromosomal stages. Heterochromatin can be further distinguished into two types:
- § Constitutive heterochromatin, which is never expressed. It is located around the centromere and usually contains repetitive sequences.
- § Facultative heterochromatin, which is sometimes expressed.

Individual chromosomes cannot be distinguished at this stage – they appear in the nucleus as a homogeneous tangled mix of DNA and protein.

Metaphase Chromatin and Division

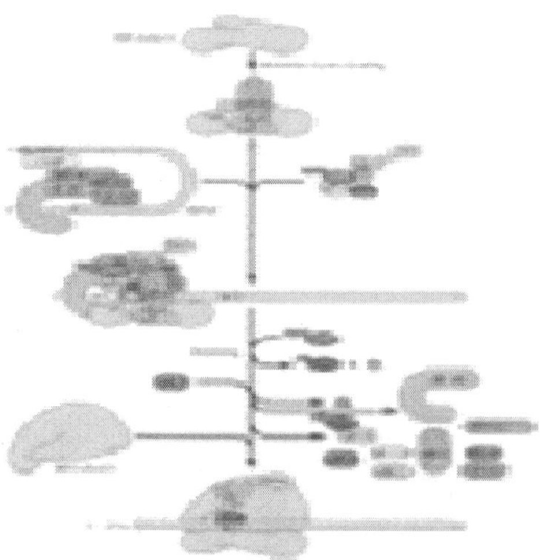

Fig. 7: Human chromosomes duringmetaphase.

In the early stages of mitosis or meiosis (cell division), the chromatin strands become more and more condensed. They cease to function as accessible genetic material (transcription stops) and become a compact transportable form.

This compact form makes the individual chromosomes visible, and they form the classic four arm structure, a pair of sister chromatids attached to each other at the centromere. The shorter arms are called p arms (from the French petit, small) and the longer arms are called q arms (q follows p in the Latin alphabet; q-g "grande"). This is the only natural context in which individual chromosomes are visible with an optical microscope.

During mitosis, microtubules grow from centrosomes located at opposite ends of the cell and also attach to the centromere at specialized structures calledkinetochores, one of which is present on each sister chromatid. A special DNA base sequence in the region of the kinetochores provides, along with special proteins, longer-lasting attachment in this region. The microtubules then pull the chromatids apart toward the centrosomes, so that each daughter cell inherits one set of chromatids. Once the cells have divided, the chromatids are uncoiled and DNA can again be transcribed. In spite of their appearance, chromosomes are structurally highly condensed, which enables these giant DNA structures to be contained within a cell nucleus.

In Humans

Chromosomes can be divided into two types—autosomes, and sex chromosomes. Certain genetic traits are linked to a person's sex and are passed on through the sex chromosomes. The autosomes contain the rest of the genetic hereditary information. All act in the same way during cell division. Human cells have 23 pairs of large linear nuclear chromosomes (22 pairs of autosomes and one pair of sex chromosomes), giving a total of 46 per cell.

In addition to these, human cells have many hundreds of copies of themitochondrial genome. Sequencing of the human genome has provided a great deal of information about each of the chromosomes. Below is a table compiling statistics for the chromosomes, based on the Sanger Institute's human genome information in the Vertebrate Genome Annotation (VEGA) database. Number of genes is an estimate as it is in part based on gene predictions. Total chromosome length is an estimate as well, based on the estimated size of unsequenced heterochromatin regions.

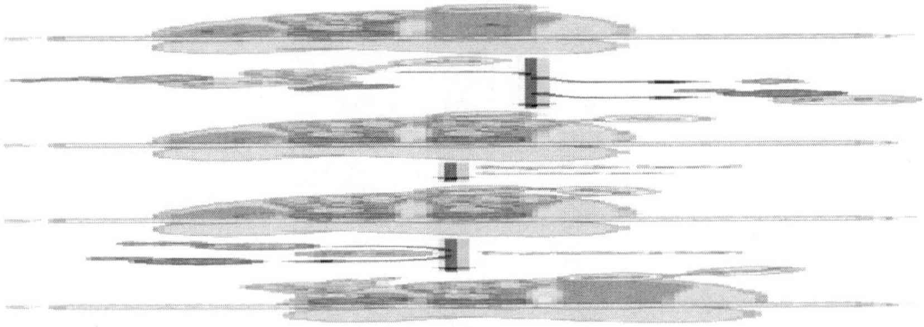

Fig. 8:

Estimated number of genes and base pairs on each human chromosome. The right-hand legend of this chart is mislabeled "Total bases". It should say "Total base pairs".

Table. 1:

Chromosome	*Genes*	*Totalbase pairs*	*Sequenced base pairs[5]*
1	4,220	247,199,719	224,999,719
2	1,491	242,751,149	237,712,649
3	1,550	199,446,827	194,704,827
4	446	191,263,063	187,297,063
5	609	180,837,866	177,702,766
6	2,281	170,896,993	167,273,993
7	2,135	158,821,424	154,952,424
8	1,106	146,274,826	142,612,826
9	1,920	140,442,298	120,312,298
10	1,793	135,374,737	131,624,737
11	379	134,452,384	131,130,853
12	1,430	132,289,534	130,303,534
13	924	114,127,980	95,559,980
14	1,347	106,360,585	88,290,585
15	921	100,338,915	81,341,915
16	909	88,822,254	78,884,754
17	1,672	78,654,742	77,800,220
18	519	76,117,153	74,656,155
19	1,555	63,806,651	55,785,651
20	1,008	62,435,965	59,505,254
21	578	46,944,323	34,171,998
22	1,092	49,528,953	34,893,953

X (sex chromosome)			
1,846	154,913,754	151,058,754	
Y (sex chromosome)	454	57,741,652	25,121,652
Total	32,185	3,079,843,747	2,857,698,560

In Prokaryotes

The prokaryotes – bacteria and archaea – typically have a single circular chromosome, but many variations exist. Most bacteria have a single circular chromosome that can range in size from only 160,000 base pairs in the endosymbiotic bacterium Candidatus Carsonella ruddii, to 12,200,000 base pairs in the soil-dwelling bacterium Sorangium cellulosum. Spirochaetes of the genus Borreliaare a notable exception to this arrangement, with bacteria such as Borrelia burgdorferi, the cause of Lyme disease, containing a single linear chromosome.

Structure in Sequences

Prokaryotic chromosomes have less sequence-based structure than eukaryotes. Bacteria typically have a single point (the origin of replication) from which replication starts, whereas some archaea contain multiple replication origins. The genes in prokaryotes are often organized in operons, and do not usually contain introns, unlike eukaryotes.

DNA Packaging

Prokaryotes do not possess nuclei. Instead, their DNA is organized into a structure called the nucleoid. The nucleoid is a distinct structure and occupies a defined region of the bacterial cell. This structure is, however, dynamic and is maintained and remodeled by the actions of a range of histone-like proteins, which associate with the bacterial chromosome. In archaea, the DNA in chromosomes is even more organized, with the DNA packaged within structures similar to eukaryotic nucleosomes.

Bacterial chromosomes tend to be tethered to the plasma membrane of the bacteria. In molecular biology application, this allows for its isolation from plasmid DNA by centrifugation of lysed bacteria and pelleting of the membranes (and the attached DNA).

Prokaryotic chromosomes and plasmids are, like eukaryotic DNA, generally supercoiled. The DNA must first be released into its relaxed state for access for transcription, regulation, and replication.

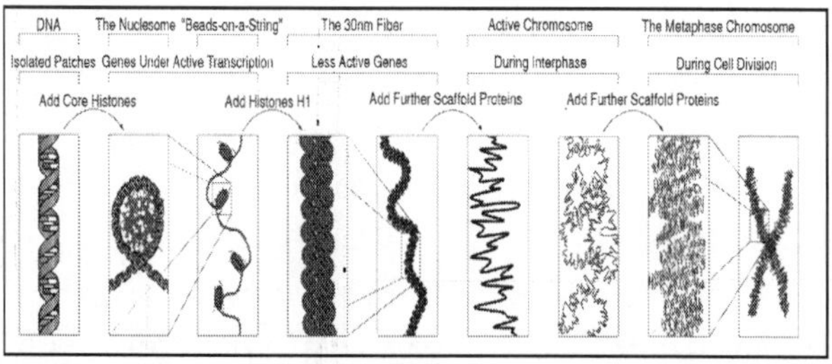

Fig. 9: The major structures in DNA compaction; DNA, the nucleosome, the 10nm "beads-on-a-string" fibre, the 30nm fibre and the metaphase chromosome.

Number of Chromosomes in Various Organisms

Eukaryotes

These tables give the total number of chromosomes (including sex chromosomes) in a cell nucleus. For example, human cells are diploid and have 22 different types of autosome, each present as two copies, and two sex chromosomes. This gives 46 chromosomes in total. Other organisms have more than two copies of their chromosomes, such as bread wheat, which is hexaploid and has six copies of seven different chromosomes – 42 chromosomes in total.

Aparticular eukaryotic species all have the same number of nuclear chromosomes. Other eukaryotic chromosomes, i.e., mitochondrial and plasmid-like small chromosomes, are much more variable in number, and there may be thousands of copies per cell.

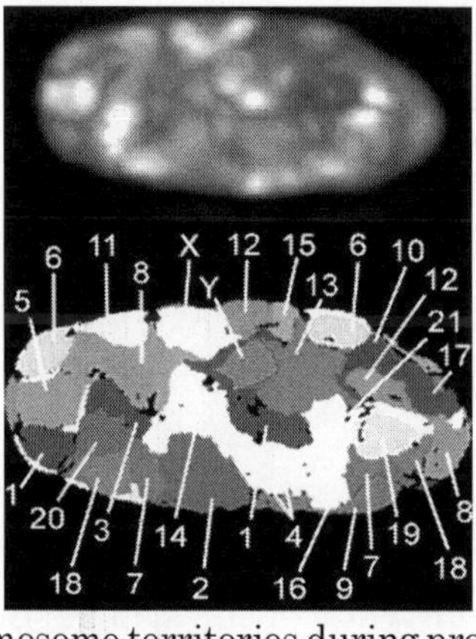

Fig. 10: The 23 human chromosome territories during prometaphase infibroblast cells.

Asexually reproducing species have one set of chromosomes, which are the same in all body cells. However, asexual species can be either haploid or diploid.

Sexually reproducing species have somatic cells (body cells), which are diploid [2n] having two sets of chromosomes, one from the mother and one from the father. Gametes, reproductive cells, are haploid [n]: They have one set of chromosomes. Gametes are produced by meiosis of a diploid germ line cell.

During meiosis, the matching chromosomes of father and mother can exchange small parts of themselves (crossover), and thus create new chromosomes that are not inherited solely from either parent. When a male and a female gamete merge (fertilization), a new diploid organism is formed.

Some animal and plant species are polyploid [Xn]: They have more than two sets of homologous chromosomes. Plants important in agriculture such as tobaccoor wheat are often polyploid, compared to their ancestral species.

Wheat has a haploid number of seven chromosomes, still seen in some cultivars as well as the wild progenitors. The more-common pasta and bread wheats are polyploid, having 28 (tetraploid) and 42 (hexaploid) chromosomes, compared to the 14 (diploid) chromosomes in the wild wheat.

Prokaryotes

Prokaryote species generally have one copy of each major chromosome, but most cells can easily survive with multiple copies. For example, Buchnera, asymbiont of aphids has multiple copies of its chromosome, ranging from 10–400 copies per cell.

However, in some large bacteria, such as Epulopiscium fishelsoni up to 100,000 copies of the chromosome can be present. Plasmids and plasmid-like small chromosomes are, as in eukaryotes, very variable in copy number. The number of plasmids in the cell is almost entirely determined by the rate of division of the plasmid – fast division causes high copy number.

Karyotype

In general, the karyotype is the characteristic chromosome complement of a eukaryote species. The preparation and study of karyotypes is part ofcytogenetics.

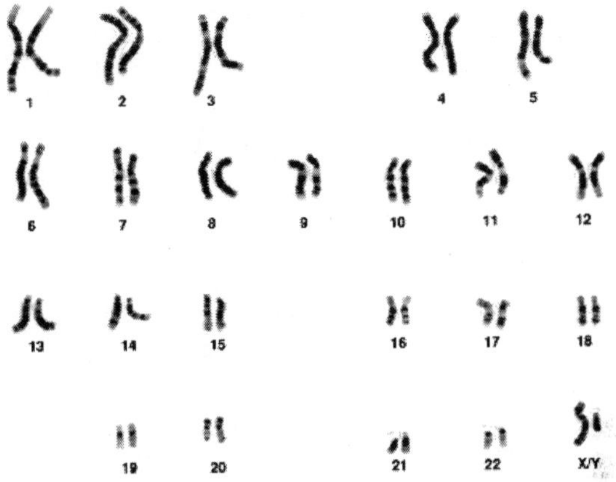

Fig. 11: Karyogram of a human male

Although the replication and transcription of DNA is highly standardized in eukaryotes, the same cannot be said for their karyotypes, which are often highly variable. There may be variation between species in chromosome number and in detailed organization. In some cases, there is significant variation within species. Often there is:

1. variation between the two sexes
2. variation between the germ-line and soma (between gametes and the rest of the body)
3. variation between members of a population, due to balanced genetic polymorphism
4. geographical variation between races
5. mosaics or otherwise abnormal individuals.

Also, variation in karyotype may occur during development from the fertilised egg.

The technique of determining the karyotype is usually called karyotyping. Cells can be locked part-way through division (in metaphase) in vitro (in a reaction vial) with colchicine. These cells are then stained, photographed, and arranged into a karyogram, with the set of chromosomes arranged, autosomes in order of length, and sex chromosomes (here X/Y) at the end: Fig. 10.

Like many sexually reproducing species, humans have special gonosomes (sex chromosomes, in contrast to autosomes). These are XX in females and XY in males.

Historical Note

Investigation into the human karyotype took many years to settle the most basic question: How many chromosomes does a normal diploid human cell contain? In 1912, Hans von Winiwarter reported 47 chromosomes in spermatogonia and 48 in oogonia, concluding an XX/XO sex determination mechanism.

Painter in 1922 was not certain whether the diploid number of man is 46 or 48, at first favouring 46. He revised his opinion later from 46 to 48, and he correctly insisted on humans having an XX/XY system.

New techniques were needed to definitively solve the problem:

1. Using cells in culture
2. Pretreating cells in a hypotonic solution, which swells them and spreads the chromosomes
3. Arresting mitosis in metaphase by a solution of colchicine
4. Squashing the preparation on the slide forcing the chromosomes into a single plane
5. Cutting up a photomicrograph and arranging the result into an indisputable karyogram.

It took until 1954 before the human diploid number was confirmed as 46. Considering the techniques of Winiwarter and Painter, their results were quite remarkable. Chimpanzees (the closest living relatives to modern humans) have 48 chromosomes.

Aberrations

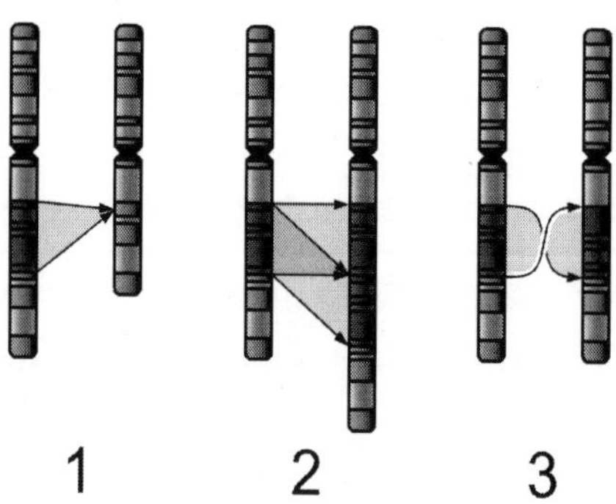

Fig. 11:

The three major single chromosome mutations; deletion (1), duplication (2) and inversion (3).

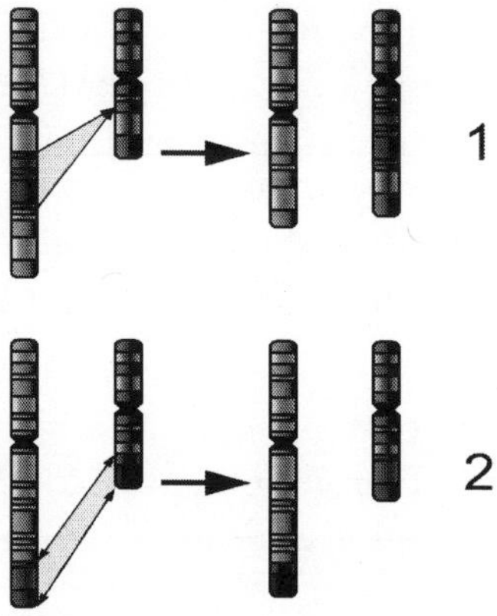

Fig. 12:

The two major two-chromosome mutations; insertion (1) and translocation (2).

Fig. 13: In Down syndrome, there are three copies of chromosome 21

Chromosomal aberrations are disruptions in the normal chromosomal content of a cell and are a major cause of genetic conditions in humans, such as Down syndrome, although most aberrations have little to no effect. Some chromosome abnormalities do not cause disease in carriers, such as translocations, orchromosomal inversions, although they may lead to a higher chance of birthing a child with a chromosome disorder. Abnormal numbers of chromosomes or chromosome sets, called aneuploidy, may be lethal or may give rise to genetic disorders. Genetic counseling is offered for families that may carry a chromosome rearrangement.

The gain or loss of DNA from chromosomes can lead to a variety of genetic disorders. Human examples include:

§ Cri du chat, which is caused by the deletion of part of the short arm of chromosome 5. "Cri du chat" means "cry of the cat" in French; the condition was so-named because affected babies make high-pitched cries that sound like those of a cat. Affected individuals have wide-set eyes, a small head and jaw, moderate to severe mental health issues, and are very short.

§ Down syndrome, the most common trisomy, usually caused by an extra copy of chromosome 21 (trisomy 21). Characteristics include decreased muscle tone, stockier build, asymmetrical skull, slanting eyes and mild to moderate developmental disability.

§ Edwards syndrome, or trisomy-18, the second-most-common trisomy. Symptoms include motor retardation, developmental disability and numerous congenital anomalies causing serious health problems. Ninety percent of those affected die in infancy; however, those that live past their first birthday are usually quite healthy thereafter. They have characteristic clenched hands and overlapping fingers.

§ Isodicentric 15, also called idic(15), partial tetrasomy 15q, or inverted duplication 15 (inv dup 15).

§ Jacobsen syndrome, which is very rare. It is also called the terminal 11q deletion disorder. Those affected have normal intelligence or mild developmental disability, with poor expressive language skills. Most have a bleeding disorder called Paris-Trousseau syndrome.

§ Klinefelter's syndrome (XXY). Men with Klinefelter syndrome are usually sterile, and tend to be taller and have longer arms and legs than their peers. Boys with the syndrome are often shy and quiet, and have a higher incidence of speech delay and dyslexia. Without testosterone treatment, some may developgynecomastia during puberty.

§ Patau Syndrome, also called D-Syndrome or trisomy-13. Symptoms are somewhat similar to those of trisomy-18, without the characteristic folded hand.

§ Small supernumerary marker chromosome. This means there is an extra, abnormal chromosome. Features depend on the origin of the extra genetic material. Cat-eye syndrome and isodicentric chromosome 15 syndrome (or Idic15) are both caused by a supernumerary marker chromosome, as isPallister-Killian syndrome.

§ Triple-X syndrome (XXX). XXX girls tend to be tall and thin and have a higher incidence of dyslexia.

§ Turner syndrome (X instead of XX or XY). In Turner syndrome, female sexual characteristics are present but underdeveloped. Females with Turner syndrome often have a short stature, low hairline, abnormal eye features and bone development and a "caved-in" appearance to the chest.

§ XYY syndrome. XYY boys are usually taller than their siblings. Like XXY boys and XXX girls, they are more likely to have learning difficulties.

§ Wolf-Hirschhorn syndrome, which is caused by partial deletion of the short arm of chromosome 4. It is characterized by severe growth retardation and severe to profound mental health issues.

DNA

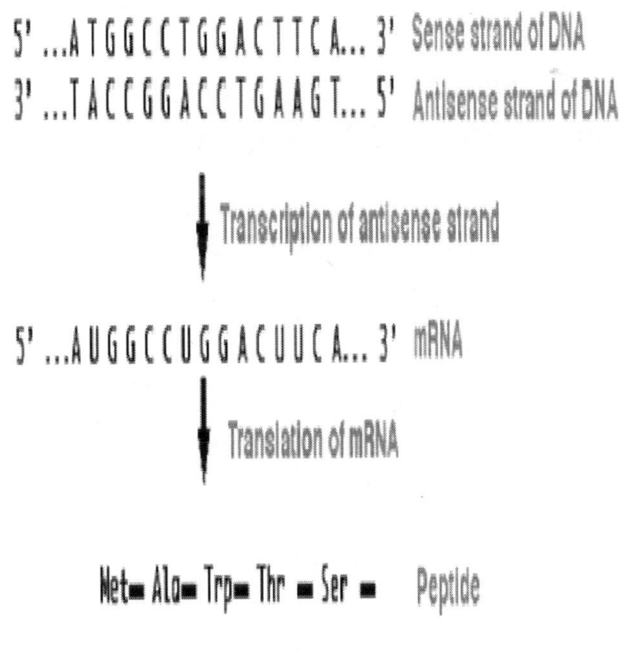

Fig. 14:

The structure of the DNA double helix. The atoms in the structure are colour coded by element and the detailed structure of two base pairs is shown in the bottom right.

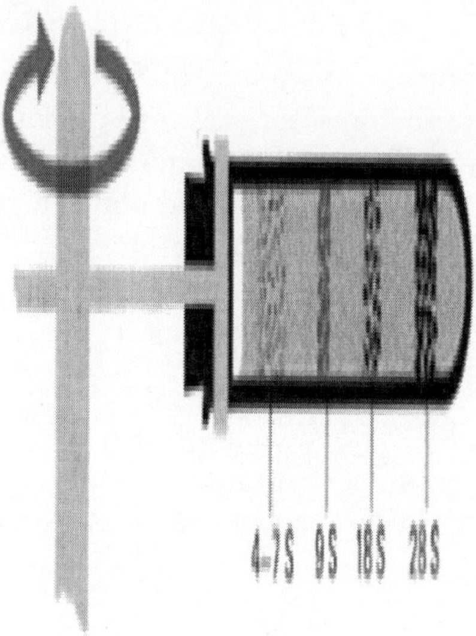

Fig. 15: The structure of part of a DNAdouble helix

Deoxyribonucleic acid DNA is a nucleic acid that contains the genetic instructions used in the development and functioning of all known living organisms (with the exception of RNA viruses). The DNA segments that carry this genetic information are called genes, but other DNA sequences have structural purposes, or are involved in regulating the use of this genetic information. Along with RNA and proteins, DNA is one of the three major macromolecules that are essential for all known forms of life.

DNA consists of two long polymers of simple units called nucleotides, with backbones made of sugars and phosphate groups joined byester bonds. These two strands run in opposite directions to each other and are therefore anti-parallel. Attached to each sugar is one of four types of molecules called nucleobases (informally, bases). It is the sequence of these four nucleobases along the backbone that encodes information. This information is read using the genetic code, which specifies the sequence of the amino acids within proteins. The code is read by copying stretches of DNA into the related nucleic acid RNA in a process called transcription.

Within cells DNA is organized into long structures called chromosomes. During cell division these chromosomes are duplicated in the process of DNA replication, providing each cell its own complete set of chromosomes. Eukaryotic organisms (animals, plants, fungi, andprotists) store most of their DNA inside the cell nucleus and some of their DNA in organelles, such as mitochondria or chloroplasts. In contrast, prokaryotes (bacteria and archaea) store their DNA only in the cytoplasm. Within the chromosomes, chromatin proteins such as histones compact and organize DNA. These compact structures guide the interactions between DNA and other proteins, helping control which parts of the DNA are transcribed.

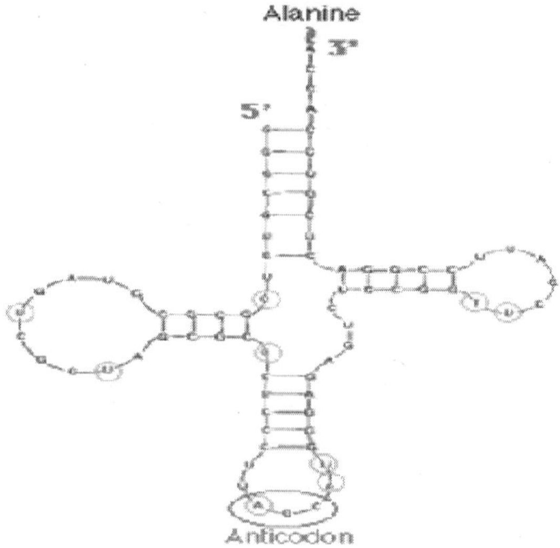

Fig. 16: Chemical structure of DNA. Hydrogen bonds shown as dotted lines.

DNA is a long polymer made from repeating units called nucleotides. As first discovered by James D. Watson and Francis Crick, the structure of DNA of all species comprises two helical chains each coiled round the same axis, and each with a pitch of 34 Ångströms(3.4 nanometres) and a radius of 10 Ångströms (1.0 nanometres). According to another study, when measured in a particular solution, the DNA chain measured 22 to 26 Ångströms wide (2.2 to 2.6 nanometres), and one nucleotide unit measured 3.3 Å (0.33 nm) long. Although each individual repeating unit is very small, DNA polymers can be very large molecules containing millions of nucleotides. For instance, the largest human chromosome, chromosome number 1, is approximately 220 million base pairs long.

In living organisms DNA does not usually exist as a single molecule, but instead as a pair of molecules that are held tightly together. These two long strands entwine like vines, in the shape of a double helix. The nucleotide repeats contain both the segment of the backbone of the molecule, which holds the chain together, and a nucleobase, which interacts with the other DNA strand in the helix. A nucleobase linked to a sugar is called a nucleoside and a base linked to a sugar and one or more phosphate groups is called a nucleotide. Polymers comprising multiple linked nucleotides (as in DNA) are called a polynucleotide.

The backbone of the DNA strand is made from alternating phosphate and sugar residues. The sugar in DNA is 2-deoxyribose, which is apentose (five-carbon) sugar. The sugars are joined together by phosphate groups that form phosphodiester bonds between the third and fifth carbon atoms of adjacent sugar rings. These asymmetric bonds mean a strand of DNA has a direction. In a double helix the direction of the nucleotides in one strand is opposite to their direction in the other strand: the strands are antiparallel. The asymmetric ends of DNA strands are called the 52 (five prime) and 32 (three prime) ends, with the 5' end having a terminal phosphate group and the 3' end a terminal hydroxyl group. One major difference between DNA and RNA is the sugar, with the 2-deoxyribose in DNA being replaced by the alternative pentose sugar ribosein RNA.

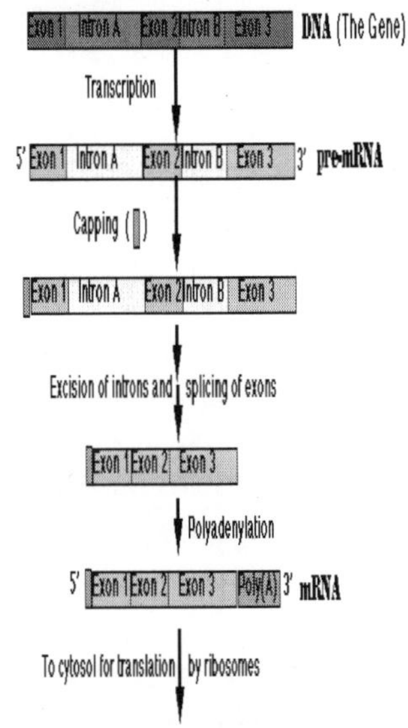

Fig. 17: A section of DNA. The bases lie horizontally between the two spiraling strands. Animated version at File:DNA orbit animated.gif.

The DNA double helix is stabilized primarily by two forces: hydrogen bonds between nucleotides and base-stacking interactions among thearomatic nucleobases. In the aqueous environment of the cell, the conjugated ∂ bonds of nucleotide bases align perpendicular to the axis of the DNA molecule, minimizing their interaction with the solvation shell and therefore, the Gibbs free energy. The four bases found in DNA areadenine (abbreviated A), cytosine (C), guanine (G) and thymine (T). These four bases are attached to the sugar/ phosphate to form the complete nucleotide, as shown for adenosine monophosphate.

The nucleobases are classified into two types: the purines, A and G, being fused five- and six-membered heterocyclic compounds, and the pyrimidines, the six-membered rings C and T. A fifth pyrimidine nucleobase, uracil (U), usually takes the place of thymine in RNA and differs from thymine by lacking a methyl group on its ring. Uracil is not usually found in DNA, occurring only as a breakdown product of cytosine. In addition to RNA and DNA a large number of artificial nucleic acid analogues have also been created to study the proprieties of nucleic acids, or for use in biotechnology.

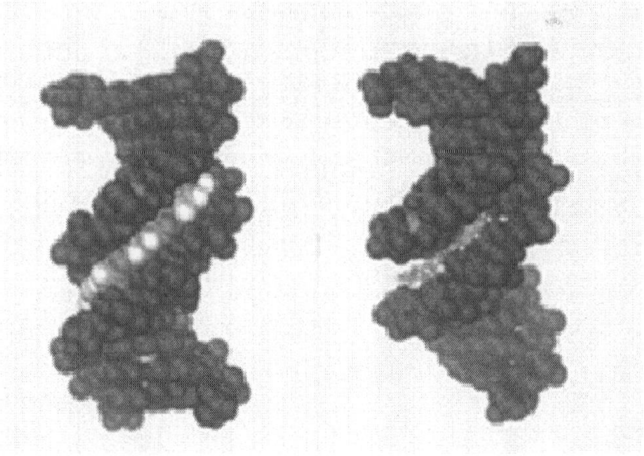

Fig. 18: Major and minor grooves of DNA. Minor groove is a binding site for the dye Hoechst 33258.

Grooves

Twin helical strands form the DNA backbone. Another double helix may be found by tracing the spaces, or grooves, between the strands. These voids are adjacent to the base pairs and may provide a binding site. As the strands are not directly opposite each other, the grooves are unequally sized. One groove, the major groove, is 22 Å wide and the other, the minor groove, is 12 Å wide. The narrowness of the minor groove means that the edges of the bases are more accessible in the major groove. As a result, proteins like transcription factors that can bind to specific sequences in double-stranded DNA usually make contacts to the sides of the bases exposed in the major groove. This situation varies in unusual conformations of DNA within the cell, but the major and minor grooves are always named to reflect the differences in size that would be seen if the DNA is twisted back into the ordinary B form.

Base Pairing

In a DNA double helix, each type of nucleobase on one strand normally interacts with just one type of nucleobase on the other strand. This is called complementarybase pairing. Here, purines form hydrogen bonds to pyrimidines, with A bonding only to T, and C bonding only to G. This arrangement of two nucleotides binding together across the double helix is called a base pair. As hydrogen bonds are not covalent, they can be broken and rejoined relatively easily. The two strands of DNA in a double helix can therefore be pulled apart like a zipper, either by a mechanical force or high temperature. As a result of this complementarity, all the information in the double-stranded sequence of a DNA helix is duplicated on each strand, which is vital in DNA replication. Indeed, this reversible and specific interaction between complementary base pairs is critical for all the functions of DNA in living organisms.

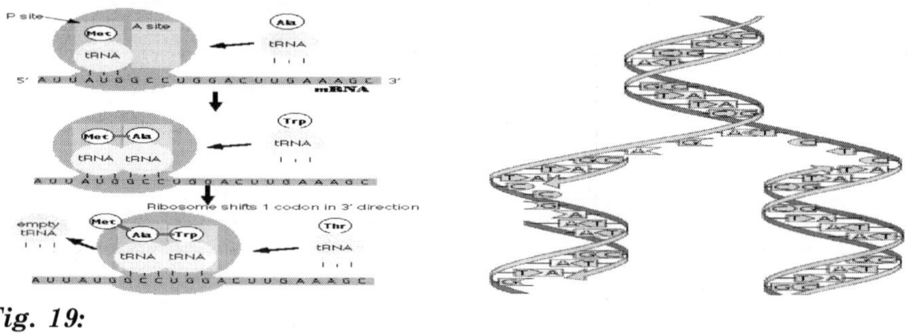

Fig. 19:

Top, a GC base pair with three hydrogen bonds. Bottom, an AT base pair with two hydrogen bonds. Non-covalent hydrogen bonds between the pairs are shown as dashed lines.

The two types of base pairs form different numbers of hydrogen bonds, AT forming two hydrogen bonds, and GC forming three hydrogen bonds. DNA with high GC-content is more stable than DNA with low GC-content. Although it is often stated that this is due to the added stability of an additional hydrogen bond, this is incorrect. DNA with high GC-content is more stable due to intra-strand base stacking interactions.

As noted above, most DNA molecules are actually two polymer strands, bound together in a helical fashion by noncovalent bonds; this double stranded structure (dsDNA) is maintained largely by the intrastrand base stacking interactions, which are strongest for G,C stacks. The two strands can come apart – a process known as melting – to form two ss DNA molecules.

Melting occurs when conditions favor ssDNA; such conditions are high temperature, low salt and high pH (low pH also melts DNA, but since DNA is unstable due to acid depurination, low pH is rarely used). The stability of the dsDNA form depends not only on the GC-content (% G,C basepairs) but also on sequence (since stacking is sequence specific) and also length (longer molecules are more stable).

The stability can be measured in various ways; a common way is the "melting temperature", which is the temperature at which 50% of the ds molecules are converted to ss molecules; melting temperature is dependent on ionic strength and the concentration of DNA.

As a result, it is both the percentage of GC base pairs and the overall length of a DNA double helix that determine the strength of the association between the two strands of DNA. Long DNA helices with a high GC-content have stronger-interacting strands, while short helices with high AT content have weaker-interacting strands.

In biology, parts of the DNA double helix that need to separate easily, such as the TATAAT Pribnow box in some promoters, tend to have a high AT content, making the strands easier to pull apart.

In the laboratory, the strength of this interaction can be measured by finding the temperature required to break the hydrogen bonds, their melting temperature (also called T_m value). When all the base pairs in a DNA double helix melt, the strands separate and exist in solution as two entirely independent molecules.

These single-stranded DNA molecules (ssDNA) have no single common shape, but some conformations are more stable than others.

Sense and Antisense

A DNA sequence is called "sense" if its sequence is the same as that of a messenger RNA copy that is translated into protein. The sequence on the opposite strand is called the "antisense" sequence. Both sense and antisense sequences can exist on different parts of the same strand of DNA (i.e. both strands contain both sense and antisense sequences). In both prokaryotes and eukaryotes, antisense RNA sequences are produced, but the functions of these RNAs are not entirely clear. One proposal is that antisense RNAs are involved in regulating gene expression through RNA-RNA base pairing.

A few DNA sequences in prokaryotes and eukaryotes, and more in plasmids and viruses, blur the distinction between sense and antisense strands by having overlapping genes. In these cases, some DNA sequences do double duty, encoding one protein when read along one strand, and a second protein when read in the opposite direction along the other strand. In bacteria, this overlap may be involved in the regulation of gene transcription, while in viruses, overlapping genes increase the amount of information that can be encoded within the small viral genome.

Supercoiling

DNA can be twisted like a rope in a process called DNA supercoiling. With DNA in its "relaxed" state, a strand usually circles the axis of the double helix once every 10.4 base pairs, but if the DNA is twisted the strands become more tightly or more loosely wound.

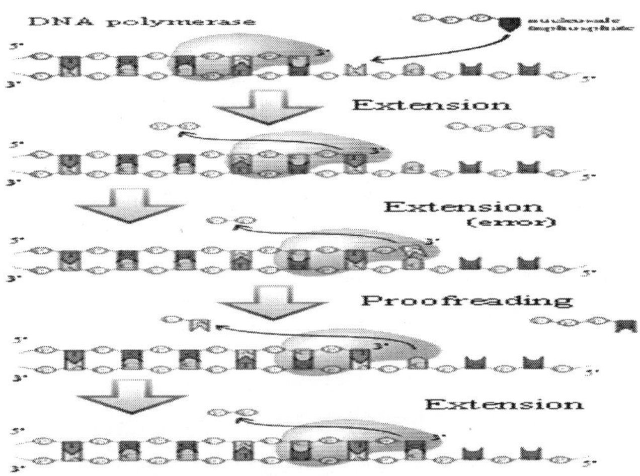

Fig. 20: From left to right, the structures of A, B and Z DNA

If the DNA is twisted in the direction of the helix, this is positive supercoiling, and the bases are held more tightly together. If they are twisted in the opposite direction, this is negative supercoiling, and the bases come apart more easily. In nature, most DNA has slight negative supercoiling that is introduced by enzymes calledtopoisomerases. These enzymes are also needed to relieve the twisting stresses introduced into DNA strands during processes such as transcription and DNA replication.

Alternate DNA Structures

DNA exists in many possible conformations that include A-DNA, B-DNA, and Z-DNA forms, although, only B-DNA and Z-DNA have been directly observed in functional organisms. The conformation that DNA adopts depends on the hydration level, DNA sequence, the amount and direction of supercoiling, chemical modifications of the bases, the type and concentration of metal ions, as well as the presence of polyamines in solution.

The first published reports of A-DNA X-ray diffraction patterns— and also B-DNA used analyses based on Patterson transforms that provided only a limited amount of structural information for oriented fibers of DNA. An alternate analysis was then proposed by Wilkins et al., in 1953, for the in vivo B-DNA X-ray diffraction/scattering patterns of highly hydrated DNA fibers in terms of squares of Bessel functions. In the same journal, James D. Watson and Francis Crick presented their molecular modeling analysis of the DNA X-ray diffraction patterns to suggest that the structure was a double-helix.

Although the 'B-DNA form' is most common under the conditions found in cells, it is not a well-defined conformation but a family of related DNA conformations that occur at the high hydration levels present in living cells. Their corresponding X-ray diffraction and scattering patterns are characteristic of molecular paracrystals with a significant degree of disorder.

Compared to B-DNA, the A-DNA form is a wider right-handed spiral, with a shallow, wide minor groove and a narrower, deeper major groove. The A form occurs under non-physiological conditions in partially dehydrated samples of DNA, while in the cell it may be produced in hybrid pairings of DNA and RNA strands, as well as in enzyme-DNA complexes. Segments of DNA where the bases have been chemically modified by methylation may undergo a larger change in conformation and adopt the Z form. Here, the strands turn about the helical axis in a left-handed spiral, the opposite of the more common B form. These unusual structures can be recognized by specific Z-DNA binding proteins and may be involved in the regulation of transcription.

Alternate DNA Chemistry

For a number of years exobiologists have proposed the existence of a shadow biosphere, a postulated microbial biosphere of Earth that uses radically different biochemical and molecular processes than currently known life. One of the proposals was the existence of lifeforms that use arsenic instead of phosphorus in DNA.

A December 2010 NASA press conference stated that the bacterium GFAJ-1, which has evolved in an arsenic-rich environment, is the first terrestrial lifeform found which may have this ability. The bacterium was found in Mono Lake, east of Yosemite National Park. GFAJ-1 is a rod-shaped extremophile bacterium in the family Halomonadaceae that, when starved of phosphorus, may be capable of incorporating the usually poisonous element arsenic in its DNA.

This discovery may lend weight to the long-standing idea that extraterrestrial life could have a different chemical makeup from life on Earth. The research was carried out by a team led by Felisa Wolfe-Simon, a geomicrobiologist and geobiochemist, a Postdoctoral Fellow of the NASA Astrobiology Institute with Arizona State University. This finding has, however, faced strong criticism from the scientific community; scientists have argued that there is no evidence that arsenic is actually incorporated into biomolecules.Independent conformation of this finding has also not yet been possible.

Quadruplex Structures

At the ends of the linear chromosomes are specialized regions of DNA called telomeres. The main function of these regions is to allow the cell to replicate chromosome ends using the enzymetelomerase, as the enzymes that normally replicate DNA cannot copy the extreme 32 ends of chromosomes. These specialized chromosome caps also help protect the DNA ends, and stop theDNA repair systems in the cell from treating them as damage to be corrected. In human cells, telomeres are usually lengths of single-stranded DNA containing several thousand repeats of a simple TTAGGG sequence.

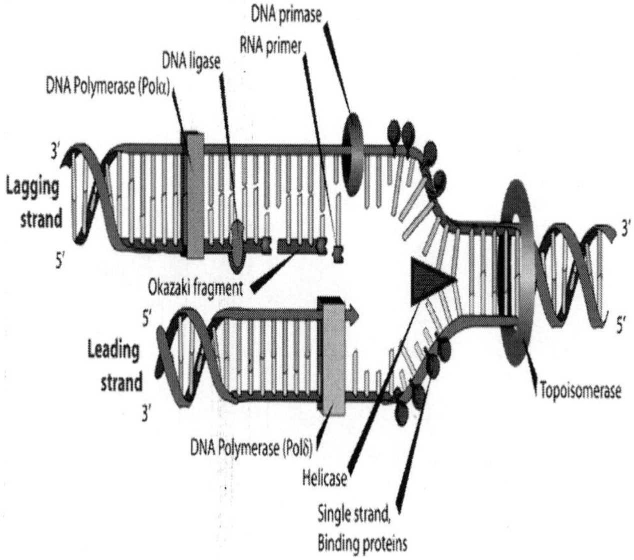

Fig. 21: DNA quadruplex formed by telomererepeats. The looped conformation of the DNA backbone is very different from the typical DNA helix.

These guanine-rich sequences may stabilize chromosome ends by forming structures of stacked sets of four-base units, rather than the usual base pairs found in other DNA molecules. Here, four guanine bases form a flat plate and these flat four-base units then stack on top of each other, to form a stable G-quadruplex structure.

These structures are stabilized by hydrogen bonding between the edges of the bases and chelation of a metal ion in the centre of each four-base unit. Other structures can also be formed, with the central set of four bases coming from either a single strand folded around the bases, or several different parallel strands, each contributing one base to the central structure.

In addition to these stacked structures, telomeres also form large loop structures called telomere loops, or T-loops. Here, the single-stranded DNA curls around in a long circle stabilized by telomere-binding proteins.

At the very end of the T-loop, the single-stranded telomere DNA is held onto a region of double-stranded DNA by the telomere strand disrupting the double-helical DNA and base pairing to one of the two strands. This triple-stranded structure is called a displacement loop or D-loop.

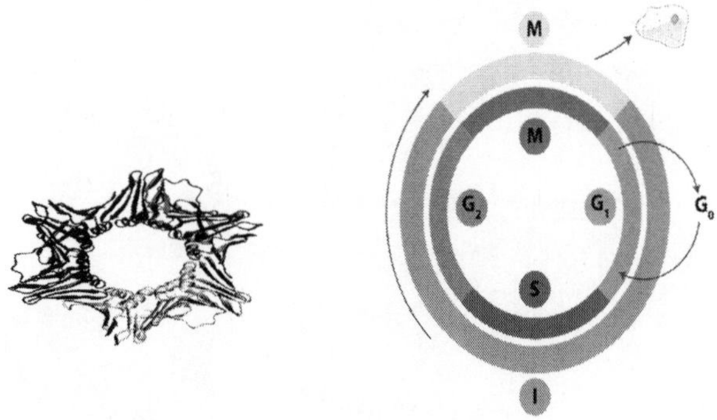

Fig. 22: Branched DNA can form networks containing multiple branches.

Branched DNA

In DNA fraying occurs when non-complementary regions exist at the end of an otherwise complementary double-strand of DNA. However, branched DNA can occur if a third strand of DNA is introduced and contains adjoining regions able to hybridize with the frayed regions of the pre-existing double-strand. Although the simplest example of branched DNA involves only three strands of DNA, complexes involving additional strands and multiple branches are also possible.Branched DNA can be used in nanotechnology to construct geometric shapes.

Vibration

DNA may carry out low-frequency collective motion as observed by the Raman spectroscopy and analyzed with a quasi-continuum model.

Chemical Modifications

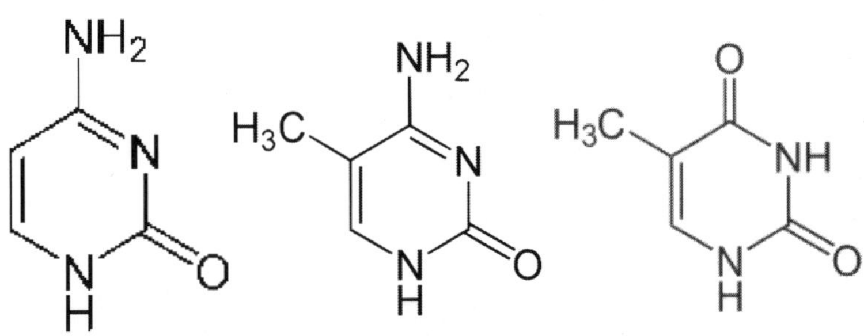

Fig. 23: Structure of cytosine with and without the 5-methyl group.Deamination converts 5-methylcytosine into thymine.

Base Modifications

The expression of genes is influenced by how the DNA is packaged in chromosomes, in a structure called chromatin. Base modifications can be involved in packaging, with regions that have low or no gene expression usually containing high levels of methylation of cytosine bases. For example, cytosine methylation, produces 5-methylcytosine, which is important for X-chromosome inactivation. The average level of methylation varies between organisms – the worm Caenorhabditis elegans lacks cytosine methylation, while vertebrates have higher levels, with up to 1% of their DNA containing 5-methylcytosine. Despite the importance of 5-methylcytosine, it can deaminate to leave a thymine base, so methylated cytosines are particularly prone to mutations. Other base modifications include adenine methylation in bacteria, the presence of 5-hydroxymethylcytosine in the brain, and the glycosylation of uracil to produce the "J-base" in kinetoplastids.

Damage

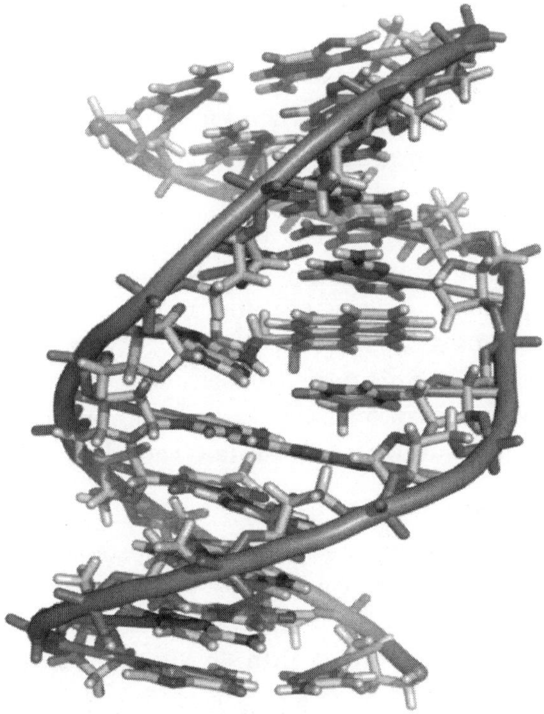

Fig. 24: A covalent adduct between ametabolically activated form ofbenzo[a]pyrene, the major mutagen intobacco smoke, and DNA

DNA can be damaged by many sorts of mutagens, which change the DNA sequence. Mutagens include oxidizing agents, alkylating agents and also high-energy electromagnetic radiation such as ultraviolet light and X-rays.

The type of DNA damage produced depends on the type of mutagen. For example, UV light can damage DNA by producing thymine dimers, which are cross-links between pyrimidine bases.

On the other hand, oxidants such asfree radicals or hydrogen peroxide produce multiple forms of damage, including base modifications, particularly of guanosine, and double-strand breaks. A typical human cell contains about 150,000 bases that have suffered oxidative damage.

Of these oxidative lesions, the most dangerous are double-strand breaks, as these are difficult to repair and can produce point mutations, insertions and deletions from the DNA sequence, as well as chromosomal translocations.

Many mutagens fit into the space between two adjacent base pairs, this is called intercalation. Most intercalators are aromatic and planar molecules; examples include ethidium bromide, acridines, daunomycin, and doxorubicin. In order for an intercalator to fit between base pairs, the bases must separate, distorting the DNA strands by unwinding of the double helix.

This inhibits both transcription and DNA replication, causing toxicity and mutations. As a result, DNA intercalators may be carcinogens, and in the case of thalidomide, a teratogen. Others such as benzo[a]pyrene diol epoxide and aflatoxin form DNA adducts which induce errors in replication. Nevertheless, due to their ability to inhibit DNA transcription and replication, other similar toxins are also used in chemotherapy to inhibit rapidly growing cancer cells.

Biological Functions

DNA usually occurs as linear chromosomes in eukaryotes, and circular chromosomes in prokaryotes. The set of chromosomes in a cell makes up itsgenome; the human genome has approximately 3 billion base pairs of DNA arranged into 46 chromosomes.

The information carried by DNA is held in thesequence of pieces of DNA called genes. Transmission of genetic information in genes is achieved via complementary base pairing.

For example, in transcription, when a cell uses the information in a gene, the DNA sequence is copied into a complementary RNA sequence through the attraction between the DNA and the correct RNA nucleotides. Usually, this RNA copy is then used to make a matching protein sequence in a process called translation, which depends on the same interaction between RNA nucleotides. In alternative fashion, a cell may simply copy its genetic information in a process called DNA replication.

The details of these functions are covered in other articles; here we focus on the interactions between DNA and other molecules that mediate the function of the genome.

Genes and Genomes

Genomic DNA is tightly and orderly packed in the process called DNA condensation to fit the small available volumes of the cell. In eukaryotes, DNA is located in the cell nucleus, as well as small amounts in mitochondria and chloroplasts. In prokaryotes, the DNA is held within an irregularly shaped body in the cytoplasm called the nucleoid.

The genetic information in a genome is held within genes, and the complete set of this information in an organism is called its genotype. A gene is a unit of heredity and is a region of DNA that influences a particular characteristic in an organism.

Genes contain an open reading frame that can be transcribed, as well as regulatory sequences such as promoters and enhancers, which control the transcription of the open reading frame.

In many species, only a small fraction of the total sequence of the genome encodes protein.

For example, only about 1.5% of the human genome consists of protein-coding exons, with over 50% of human DNA consisting of non-coding repetitive sequences.

The reasons for the presence of so much noncoding DNA in eukaryotic genomes and the extraordinary differences in genome size, or C-value, among species represent a long-standing puzzle known as the "C-value enigma".

However, DNA sequences that do not code protein may still encode functional non-coding RNA molecules, which are involved in the regulation of gene expression.

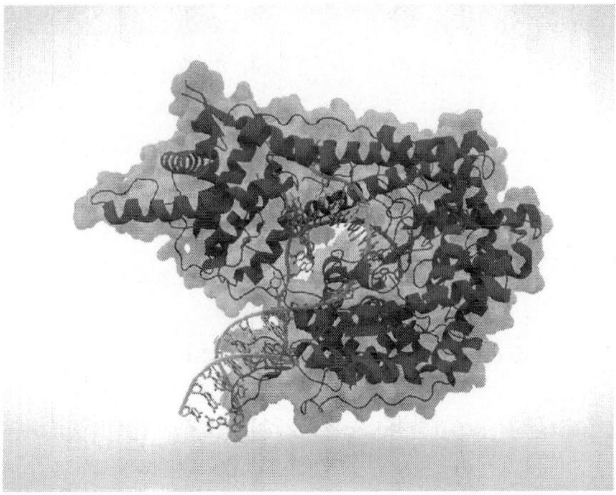

Fig. 25: T7 RNA polymerase (blue) producing a mRNA (green) from a DNA template (orange).

Some noncoding DNA sequences play structural roles in chromosomes. Telomeres and centromeres typically contain few genes, but are important for the function and stability of chromosomes. An abundant form of noncoding DNA in humans are pseudogenes, which are copies of genes that have been disabled by mutation. These sequences are usually just molecular fossils, although they can occasionally serve as raw genetic material for the creation of new genes through the process of gene duplication and divergence.

Transcription and Translation

A gene is a sequence of DNA that contains genetic information and can influence the phenotype of an organism. Within a gene, the sequence of bases along a DNA strand defines a messenger RNA sequence, which then defines one or more protein sequences. The relationship between the nucleotide sequences of genes and the amino-acid sequences of proteins is determined by the rules of translation, known collectively as the genetic code. The genetic code consists of three-letter 'words' called codons formed from a sequence of three nucleotides (e.g. ACT, CAG, TTT).

In transcription, the codons of a gene are copied into messenger RNA by RNA polymerase. This RNA copy is then decoded by a ribosome that reads the RNA sequence by base-pairing the messenger RNA to transfer RNA, which carries amino acids. Since there are 4 bases in 3-letter combinations, there are 64 possible codons (4 combinations). These encode the twenty standard amino acids, giving most amino acids more than one possible codon. There are also three 'stop' or 'nonsense' codons signifying the end of the coding region; these are the TAA, TGA and TAG codons.

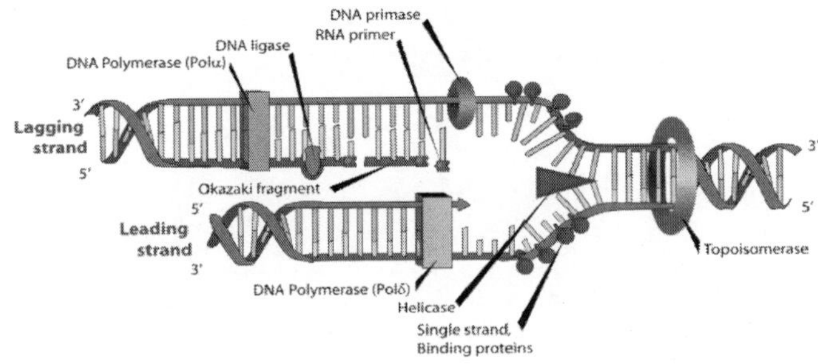

Fig. 26:

DNA replication. The double helix is unwound by a helicase and topoisomerase. Next, one DNA polymerase produces the leading strand copy. Another DNA polymerase binds to the lagging strand. This enzyme makes discontinuous segments (called Okazaki fragments) before DNA ligase joins them together.

Replication

Cell division is essential for an organism to grow, but, when a cell divides, it must replicate the DNA in its genome so that the two daughter cells have the same genetic information as their parent. The double-stranded structure of DNA provides a simple mechanism for DNA replication. Here, the two strands are separated and then each strand'scomplementary DNA sequence is recreated by an enzyme called DNA polymerase. This enzyme makes the complementary strand by finding the correct base through complementary base pairing, and bonding it onto the original strand. As DNA polymerases can only extend a DNA strand in a 52 to 32 direction, different mechanisms are used to copy the antiparallel strands of the double helix. In this way, the base on the old strand dictates which base appears on the new strand, and the cell ends up with a perfect copy of its DNA.

Interactions with Proteins

All the functions of DNA depend on interactions with proteins. These protein interactions can be non-specific, or the protein can bind specifically to a single DNA sequence. Enzymes can also bind to DNA and of these, the polymerases that copy the DNA base sequence in transcription and DNA replication are particularly important.

DNA-binding Proteins

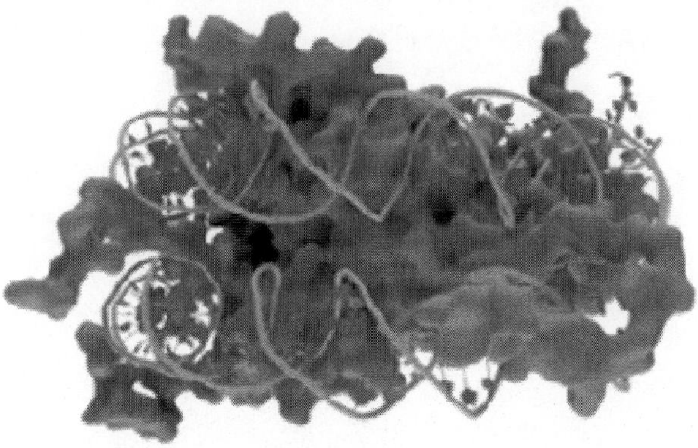

Fig. 27: Interaction of DNA (shown in orange) withhistones (shown in blue). These proteins' basic amino acids bind to the acidic phosphate groups on DNA.

Structural proteins that bind DNA are well-understood examples of non-specific DNA-protein interactions. Within chromosomes, DNA is held in complexes with structural proteins. These proteins organize the DNA into a compact structure called chromatin. In eukaryotes this structure involves DNA binding to a complex of small basic proteins called histones, while in prokaryotes multiple types of proteins are involved.

The histones form a disk-shaped complex called a nucleosome, which contains two complete turns of double-stranded DNA wrapped around its surface. These non-specific interactions are formed through basic residues in the histones making ionic bonds to the acidic sugar-phosphate backbone of the DNA, and are therefore largely independent of the base sequence. Chemical modifications of these basic amino acid residues include methylation, phosphory-lationand acetylation. These chemical changes alter the strength of the interaction between the DNA and the histones, making the DNA more or less accessible to transcription factors and changing the rate of transcription. Other non-specific DNA-binding proteins in chromatin include the high-mobility group proteins, which bind to bent or distorted DNA. These proteins are important in bending arrays of nucleosomes and arranging them into the larger structures that make up chromosomes.

A distinct group of DNA-binding proteins are the DNA-binding proteins that specifically bind single-stranded DNA. In humans, replication protein A is the best-understood member of this family and is used in processes where the double helix is separated, including

DNA replication, recombination and DNA repair. These binding proteins seem to stabilize single-stranded DNA and protect it from forming stem-loops or being degraded by nucleases.

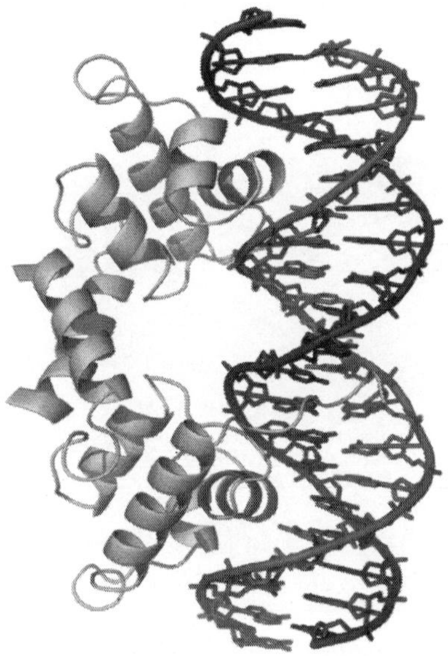

Fig. 28: The lambda repressor helix-turn-helix transcription factor bound to its DNA target

In contrast, other proteins have evolved to bind to particular DNA sequences. The most intensively studied of these are the various transcription factors, which are proteins that regulate transcription. Each transcription factor binds to one particular set of DNA sequences and activates or inhibits the transcription of genes that have these sequences close to their promoters. The transcription factors do this in two ways. Firstly, they can bind the RNA polymerase responsible for transcription, either directly or through other mediator proteins; this locates the polymerase at the promoter and allows it to begin transcription. Alternatively, transcription factors can bind enzymes that modify the histones at the promoter; this will change the accessibility of the DNA template to the polymerase.

As these DNA targets can occur throughout an organism's genome, changes in the activity of one type of transcription factor can affect thousands of genes.Consequently, these proteins are often the targets of the signal transduction processes that control responses to environmental changes or cellular differentiation and development. The specificity of these transcription factors' interactions with DNA come from the proteins making multiple contacts to the edges of the DNA bases, allowing them to "read" the DNA sequence. Most of these base-interactions are made in the major groove, where the bases are most accessible.

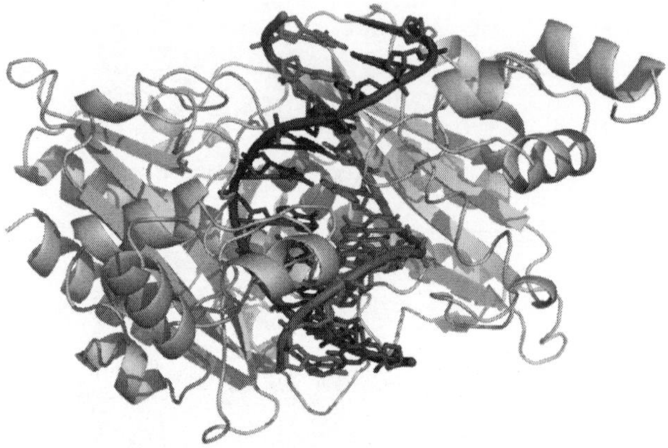

Fig. 29: The restriction enzyme EcoRV (green) in a complex with its substrate DNA

DNA-modifying Enzymes

Nucleases and Ligases

Nucleases are enzymes that cut DNA strands by catalyzing the hydrolysis of the phosphodiester bonds. Nucleases that hydrolyse nucleotides from the ends of DNA strands are called exonucleases, while endonucleases cut within strands. The most frequently used nucleases in molecular biology are the restriction endonucleases, which cut DNA at specific sequences. For instance, the EcoRV enzyme shown to the left recognizes the 6-base sequence 52 -GAT | ATC-32 and makes a cut at the vertical line. In nature, these enzymes protect bacteria against phage infection by digesting the phage DNA when it enters the bacterial cell, acting as part of the restriction modification system. In technology, these sequence-specific nucleases are used in molecular cloning and DNA fingerprinting.

Enzymes called DNA ligases can rejoin cut or broken DNA strands. Ligases are particularly important in lagging strand DNA replication, as they join together the short segments of DNA produced at the replication fork into a complete copy of the DNA template. They are also used in DNA repair and genetic recombination.

Topoisomerases and Helicases

Topoisomerases are enzymes with both nuclease and ligase activity. These proteins change the amount of supercoiling in DNA. Some of these enzymes work by cutting the DNA helix and allowing one section to rotate, thereby reducing its level of supercoiling; the enzyme then seals the DNA break. Other types of these enzymes are capable of cutting one DNA helix and then passing a second strand of DNA through this break, before rejoining the helix. Topoisomerases are required for many processes involving DNA, such as DNA replication and transcription.

Helicases are proteins that are a type of molecular motor. They use the chemical energy in nucleoside triphosphates, predominantly ATP, to break hydrogen bonds between bases and unwind the DNA double helix into single strands. These enzymes are essential for most processes where enzymes need to access the DNA bases.

Polymerases

Polymerases are enzymes that synthesize polynucleotide chains from nucleoside triphosphates. The sequence of their products are copies of existing polynucleotide chains – which are calledtemplates. These enzymes function by adding nucleotides onto the 32 hydroxyl group of the previous nucleotide in a DNA strand. As a consequence, all polymerases work in a 52 to 32 direction. In the active site of these enzymes, the incoming nucleoside triphosphate base-pairs to the template: this allows polymerases to accurately synthesize the complementary strand of their template. Polymerases are classified according to the type of template that they use.

In DNA replication, a DNA-dependent DNA polymerase makes a copy of a DNA sequence. Accuracy is vital in this process, so many of these polymerases have a proofreading activity. Here, the polymerase recognizes the occasional mistakes in the synthesis reaction by the lack of base pairing between the mismatched nucleotides. If a mismatch is detected, a 32 to 52 exonuclease activity is activated and the incorrect base removed. In most organisms, DNA polymerases function in a large complex called the replisome that contains multiple accessory subunits, such as the DNA clampor helicases.

RNA-dependent DNA polymerases are a specialized class of polymerases that copy the sequence of an RNA strand into DNA. They include reverse transcriptase, which is a viral enzyme involved in the infection of cells by retroviruses, and telomerase, which is required for the replication of telomeres. Telomerase is an unusual polymerase because it contains its own RNA template as part of its structure.

Transcription is carried out by a DNA-dependent RNA polymerase that copies the sequence of a DNA strand into RNA. To begin transcribing a gene, the RNA polymerase binds to a sequence of DNA called a promoter and separates the DNA strands. It then copies the gene sequence into a messenger RNA transcript until it reaches a region of DNA called the terminator, where it halts and detaches from the DNA. As with human DNA-dependent DNA polymerases, RNA polymerase II, the enzyme that transcribes most of the genes in the human genome, operates as part of a large protein complexwith multiple regulatory and accessory subunits.

Genetic Recombination

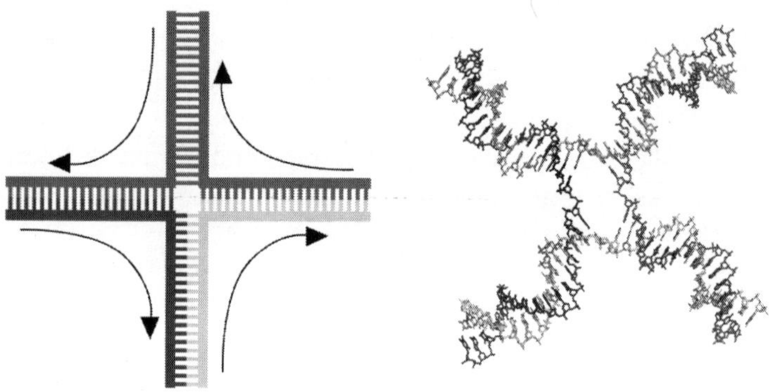

Fig. 30:

Structure of the Holliday junctionintermediate in genetic recombination. The four separate DNA strands are coloured red, blue, green and yellow.

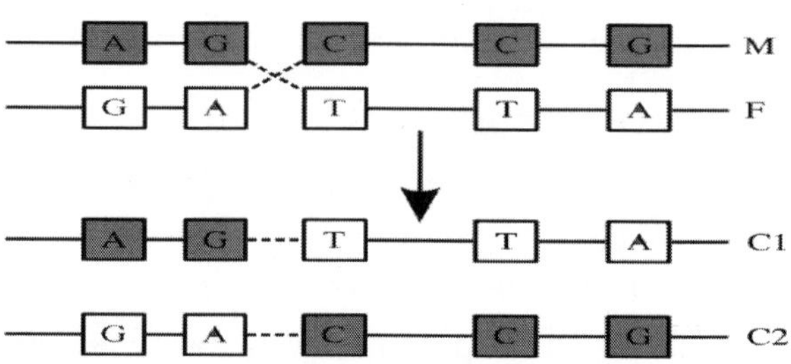

Fig. 31: Recombination involves the breakage and rejoining of two chromosomes (M and F) to produce two re-arranged chromosomes (C1 and C2).

A DNA helix usually does not interact with other segments of DNA, and in human cells the different chromosomes even occupy separate areas in the nucleus called "chromosome territories". This physical separation of different chromosomes is important for the ability of DNA to function as a stable repository for information, as one of the few times chromosomes interact is during chromosomal crossover when they recombine. Chromosomal crossover is when two DNA helices break, swap a section and then rejoin.

Recombination allows chromosomes to exchange genetic information and produces new combinations of genes, which increases the efficiency of natural selection and can be important in the rapid evolution of new proteins. Genetic recombination can also be involved in DNA repair, particularly in the cell's response to double-strand breaks.

The most common form of chromosomal crossover is homologous recombination, where the two chromosomes involved share very similar sequences. Non-homologous recombination can be damaging to cells, as it can produce chromosomal translocations and genetic abnormalities. The recombination reaction is catalyzed by enzymes known as recombinases, such as RAD51. The first step in recombination is a double-stranded break either caused by an endonuclease or damage to the DNA. A series of steps catalyzed in part by the recombinase then leads to joining of the two helices by at least one Holliday junction, in which a segment of a single strand in each helix is annealed to the complementary strand in the other helix. The Holliday junction is a tetrahedral junction structure that can be moved along the pair of chromosomes, swapping one strand for another. The recombination reaction is then halted by cleavage of the junction and re-ligation of the released DNA.

Evolution

DNA contains the genetic information that allows all modern living things to function, grow and reproduce. However, it is unclear how long in the 4-billion-year history of life DNA has performed this function, as it has been proposed that the earliest forms of life may have used RNA as their genetic material. RNA may have acted as the central part of early cell metabolism as it can both transmit genetic information and carry out catalysis as part of ribozymes. This ancient RNA world where nucleic acid would have been used for both catalysis and genetics may have influenced theevolution of the current genetic code based on four nucleotide bases. This would occur, since the number of different bases in such an organism is a trade-off between a small number of bases increasing replication accuracy and a large number of bases increasing the catalytic efficiency of ribozymes.

However, there is no direct evidence of ancient genetic systems, as recovery of DNA from most fossils is impossible. This is because DNA will survive in the environment for less than one million years and slowly degrades into short fragments in solution. Claims for older DNA have been made, most notably a report of the isolation of a viable bacterium from a salt crystal 250 million years old,but these claims are controversial.

On August 8, 2011, a report, based on NASA studies with meteorites found on Earth, was published suggesting building blocks of DNA (adenine, guanine and related organic molecules) may have been formed extraterrestrially in outer space.

Uses in Technology

Genetic Engineering

Methods have been developed to purify DNA from organisms, such as phenol-chloroform extraction, and to manipulate it in the laboratory, such as restriction digests and the polymerase chain reaction. Modern biology and biochemistry make intensive use of these techniques in recombinant DNA technology. Recombinant DNA is a man-made DNA sequence that has been assembled from other DNA sequences. They can be transformed into organisms in the form of plasmids or in the appropriate format, by using a viral vector. The genetically modified organisms produced can be used to produce products such as recombinant proteins, used in medical research, or be grown in agriculture.

Forensics

Forensic scientists can use DNA in blood, semen, skin, saliva or hair found at a crime scene to identify a matching DNA of an individual, such as a perpetrator. This process is formally termed DNA profiling, but may also be called "genetic fingerprinting". In DNA profiling, the lengths of variable sections of repetitive DNA, such as short tandem repeats and minisatellites, are compared between people. This method is usually an extremely reliable technique for identifying a matching DNA. However, identification can be complicated if the scene is contaminated with DNA from several people. DNA profiling was developed in 1984 by British geneticist Sir Alec Jeffreys, and first used in forensic science to convict Colin Pitchfork in the 1988 Enderby murders case.

People convicted of certain types of crimes may be required to provide a sample of DNA for a database. This has helped investigators solve old cases where only a DNA sample was obtained from the scene. DNA profiling can also be used to identify victims of mass casualty incidents. On the other hand, many convicted people have been released from prison on the basis of DNA techniques, which were not available when a crime had originally been committed.

Bioinformatics

Bioinformatics involves the manipulation, searching, and data mining of biological data, and this includes DNA sequence data. The development of techniques to store and search DNA sequences have led to widely applied advances in computer science, especially string searching algorithms, machine learning and database theory. String searching or matching algorithms, which find an occurrence of a sequence of letters inside a larger sequence of letters, were developed to search for specific sequences of nucleotides. The DNA sequenced may be aligned with other DNA sequences to identify homologous sequences and locate the specific mutations that make them distinct.

These techniques, especially multiple sequence alignment, are used in studying phylogeneticrelationships and protein function. Data sets representing entire genomes' worth of DNA sequences, such as those produced by the Human Genome Project, are difficult to use without the annotations that identify the locations of genes and regulatory elements on each chromosome. Regions of DNA sequence that have the characteristic patterns associated with protein- or RNA-coding genes can be identified by gene finding algorithms, which allow researchers to predict the presence of particular gene products and their possible functions in an organism even before they have been isolated experimentally. Entire genomes may also be compared which can shed light on the evolutionary history of particular organism and permit the examination of complex evolutionary events.

Chapter 7

Genome Organization and Evolution

Genomics is a discipline in genetics concerning the study of the genomes of organisms. The field includes intensive efforts to determine the entire DNA sequence of organisms and fine-scale genetic mapping efforts. The field also includes studies of intragenomic phenomena such as heterosis, epistasis, pleiotropy and other interactions between loci and alleles within the genome. In contrast, the investigation of the roles and functions of single genes is a primary focus of molecular biology or genetics and is a common topic of modern medical and biological research. Research of single genes does not fall into the definition of genomics unless the aim of this genetic, pathway, and functional information analysis is to elucidate its effect on, place in, and response to the entire genome's networks.

For the United States Environmental Protection Agency, "the term "genomics" encompasses a broader scope of scientific inquiry associated technologies than when genomics was initially considered. A genome is the sum total of all an individual organism's genes. Thus, genomics is the study of all the genes of a cell, or tissue, at the DNA (genotype), mRNA (transcriptome), or protein (proteome) levels."

History

Genomics was established by Fred Sanger when he first sequenced the complete genomes of a virus and a mitochondrion. His group established techniques of sequencing, genome mapping, data storage, and bioinformatic analyses in the 1970-1980s. A major branch of genomics is still concerned with sequencing the genomes of various organisms, but the knowledge of full genomes has created the possibility for the field of functional genomics, mainly concerned with patterns of gene expression during various conditions. The most important tools here are microarrays and bioinformatics. Study of the full set of proteins in a cell type or tissue, and the changes during various conditions, is called proteomics. A related concept is materiomics, which is defined as the study of the material properties of biological materials (e.g. hierarchical protein structures and materials, mineralized biological tissues, etc.) and their effect on the macroscopic function and failure in their biological context, linking processes, structure and properties at multiple scales through a materials science approach. The actual term 'genomics' is thought to have been coined by Dr. Tom Roderick, a geneticist at the Jackson Laboratory (Bar Harbor, ME) over beer at a meeting held in Maryland on the mapping of the human genome in 1986.

In 1972, Walter Fiers and his team at the Laboratory of Molecular Biology of the University of Ghent (Ghent, Belgium) were the first to determine the sequence of a gene: the gene for Bacteriophage MS2 coat protein. In 1976, the team determined the complete nucleotide-sequence of bacteriophage MS2-RNA. The first DNA-based genome to be sequenced in its entirety was that of bacteriophage Φ-X174; (5,368 bp), sequenced by Frederick Sanger in 1977.

The first free-living organism to be sequenced was that of *Haemophilus influenzae* (1.8 Mb) in 1995, and since then genomes are being sequenced at a rapid pace.

As of September 2007, the complete sequence was known of about 1879 viruses, 577 bacterial species and roughly 23 eukaryote organisms, of which about half are fungi. Most of the bacteria whose genomes have been completely sequenced are problematic disease-causing agents, such as *Haemophilus influenzae*. Of the other sequenced species, most were chosen because they were well-studied model organisms or promised to become good models. Yeast (*Saccharomyces cerevisiae*) has long been an important model organism for the eukaryotic cell, while the fruit fly *Drosophila melanogaster* has been a very important tool (notably in early pre-molecular genetics). The worm *Caenorhabditis elegans* is an often used simple model for multicellular organisms.

The zebrafish *Brachydanio rerio* is used for many developmental studies on the molecular level and the flower *Arabidopsis thaliana* is a model organism for flowering plants. The Japanese pufferfish (*Takifugu rubripes*) and the spotted green pufferfish (*Tetraodon nigroviridis*) are interesting because of their small and compact genomes, containing very little non-coding DNA compared to most species. The mammals dog (*Canis familiaris*), brown rat (*Rattus norvegicus*), mouse (*Mus musculus*), and chimpanzee (*Pan troglodytes*) are all important model animals in medical research.

Human Genomics

A rough draft of the human genome was completed by the Human Genome Project in early 2001, creating much fanfare. By 2007 the human sequence was declared "finished" (less than one error in 20,000 bases and all chromosomes assembled). Display of the results of the project required significant bioinformatics resources. The sequence of the human reference assembly can be explored using the UCSC Genome Browser.

Bacteriophage Genomics

Bacteriophages have played and continue to play a key role in bacterial genetics and molecular biology. Historically, they were used to define gene structure and gene regulation. Also the first genome to be sequenced was a bacteriophage. However, bacteriophage research did not lead the genomics revolution, which is clearly dominated by bacterial genomics.

Only very recently has the study of bacteriophage genomes become prominent, thereby enabling researchers to understand the mechanisms underlying phage evolution. Bacteriophage genome sequences can be obtained through direct sequencing of isolated bacteriophages, but can also be derived as part of microbial genomes. A detailed database mining of these sequences offers insights into the role of prophages in shaping the bacterial genome.

Cyanobacteria Genomics

At present there are 24 cyanobacteria for which a total genome sequence is available. 15 of these cyanobacteria come from the marine environment. These are six *Prochlorococcus* strains, seven marine *Synechococcus* strains, *Trichodesmium erythraeum* IMS101 and *Crocosphaera watsonii* WH8501. Several studies have demonstrated how these sequences could be used very successfully to infer important ecological and physiological characteristics of marine cyanobacteria.

However, there are many more genome projects currently in progress, amongst those there are further *Prochlorococcus* and marine *Synechococcus* isolates, *Acaryochloris* and *Prochloron*, the N_2-fixing filamentous cyanobacteria *Nodularia spumigena*, *Lyngbya aestuarii* and *Lyngbya majuscula*, as well as bacteriophages infecting marine cyanobaceria. Thus, the growing body of genome information can also be tapped in a more general way to address global problems by applying a comparative approach.

Some new and exciting examples of progress in this field are the identification of genes for regulatory RNAs, insights into the evolutionary origin of photosynthesis, or estimation of the contribution of horizontal gene transfer to the genomes that have been analysed.

Genomes and Proteomes

The genome of a typical bacterium comes as a single DNA molecule, which, if extended, would be about 2 mm long. (The cell itself has a diameter of about 0.001 mm). The DNA of higher organisms is organized into chromosomes—normal human cells contain 23 chromosome pairs. The total amount of genetic information per cell—the sequence of nucleotides of DNA—is very nearly constant for all members of a species.

Learning Goals

1. To know the basic sizes and organizing principles of simple and complex genomes.
2. To understand ideas of proteomics—the patterns of expression of proteins—and techniques such as microarrays used to study them.
3. To know how genomes are analysed, and the relation of gene sequenceto phenotypic features, including inherited diseases.
4. To absorb a general idea of the contents of particular genomes, and how the genomes of prokaryotes and eukaryotes differ systematically.
5. To recognize both the importance and the difficulty of deriving from a complete genome sequence the amino acid sequences of the proteins encoded, and assigning functions to these proteins.
6. To realize that published genomes record the characteristics of only a single individual, and that there is considerable variation within populations, and great variation between separated populations of organisms belonging to the same species.

7. To appreciate the power of comparative genomics to identify features responsible for differences between species—for instance, what is it that makes us human?

Gene

A gene is a unit of heredity in a living organism. It normally resides on a stretch of DNA that codes for a type of protein or for an RNA chain that has a function in the organism. All living things depend on genes, as they specify all proteins and functional RNA chains. Genes hold the information to build and maintain an organism's cells and pass genetic traits to offspring, although some organelles (e.g. mitochondria) are self-replicating and are not coded for by the organism's DNA.

A modern working definition of a gene is "*a locatable region of genomic sequence, corresponding to a unit of inheritance, which is associated with regulatory regions, transcribed regions, and or other functional sequence regions* ". Colloquial usage of the term *gene* (e.g. "good genes, "hair colour gene") may actually refer to an allele: a *gene* is the basic instruction, a sequence of nucleic acid (DNA or, in the case of certain viruses RNA), while an *allele* is one variant of that gene. Thus, when the mainstream press refers to "having" a "gene" for a specific trait, this may be incorrect. In many cases, all people would have the gene in question, but certain people will have a specific allele of that gene, which results in the trait. In the simplest case, the changes observed may be caused by a single letter of the genetic code - a single nucleotide polymorphism.

The notion of a gene is evolving with the science of genetics, which began when Gregor Mendel noticed that biological variations are inherited from parent organisms as specific, discrete traits. The biological entity responsible for defining traits was later termed a *gene*, but the biological basis for inheritance remained unknown until DNA was identified as the genetic material in the 1940s. All organisms have many genes corresponding to many different biological traits, some of which are immediately visible, such as eye colour or number of limbs, and some of which are not, such as blood type or increased risk for specific diseases, or the thousands of basic biochemical processes that comprise life.

The vast majority of living organisms encode their genes in long strands of DNA. DNA (deoxyribonucleic acid) consists of a chain made from four types of nucleotide subunits, each composed of: a five-carbon sugar (2'-deoxyribose), a phosphate group, and one of the four bases adenine, cytosine, guanine, and thymine. The most common form of DNA in a cell is in a double helix structure, in which two individual DNA strands twist around each other in a right-handed spiral. In this structure, the base pairing rules specify that guanine pairs with cytosine and adenine pairs with thymine. The base pairing between guanine and cytosine forms three hydrogen bonds, whereas the base pairing between adenine and thymine forms two hydrogen bonds. The two strands in a double helix must therefore be *complementary*, that is, their bases must align such that the adenines of one strand are paired with the thymines of the other strand, and so on.

Due to the chemical composition of the pentose residues of the bases, DNA strands have directionality. One end of a DNA polymer contains an exposed hydroxyl group on the deoxyribose; this is known as the 3' end of the molecule. The other end contains an exposed phosphate group; this is the 5' end. The directionality of DNA is vitally important to many cellular processes, since double helices are necessarily directional (a strand running 5'-3' pairs with a complementary strand running 3'-5'), and processes such as DNA replication occur in only one direction. All nucleic acid synthesis in a cell occurs in the 5'-3' direction, because new monomers are added via a dehydration reaction that uses the exposed 3' hydroxyl as a nucleophile. The expression of genes encoded in DNA begins by transcribing the gene into RNA, a second type of nucleic acid that is very similar to DNA, but whose monomers contain the sugar ribose rather than deoxyribose. RNA also contains the base uracil in place of thymine. RNA molecules are less stable than DNA and are typically single-stranded. Genes that encode proteins are composed of a series of three-nucleotide sequences called codons, which serve as the *words* in the genetic *language*. The genetic code specifies the correspondence during protein translation between codons and amino acids. The genetic code is nearly the same for all known organisms.

Physical Definitions

RNA Genes and Genomes

When proteins are manufactured, the gene is first copied into RNA as an intermediate product. In other cases, the RNA molecules are the actual functional products. For example, RNAs known as ribozymes are capable of enzymatic function, and microRNA has a regulatory role. The DNA sequences from which such RNAs are transcribed are known as RNA genes.

Some viruses store their entire genomes in the form of RNA, and contain no DNA at all. Because they use RNA to store genes, their cellular hosts may synthesize their proteins as soon as they are infected and without the delay in waiting for transcription. On the other hand, RNA retroviruses, such as HIV, require the reverse transcription of their genome from RNA into DNA before their proteins can be synthesized. In 2006, French researchers came across a puzzling example of RNA-mediated inheritance in mouse. Mice with a loss-of-function mutation in the gene Kit have white tails. Offspring of these mutants can have white tails despite having only normal Kit genes. The research team traced this effect back to mutated Kit RNA. While RNA is common as genetic storage material in viruses, in mammals in particular RNA inheritance has been observed very rarely.

Functional Structure of a Gene

All genes have regulatory regions in addition to regions that explicitly code for a protein or RNA product. A regulatory region shared by almost all genes is known as the promoter, which provides a position that is recognized by the transcription machinery when a gene is about to be transcribed and expressed. A gene can have more than one promoter, resulting in RNAs that differ in how far they extend in the 5' end. Although promoter regions have a consensus sequence that is the most common sequence at this position, some genes have "strong" promoters that bind the transcription machinery well, and others have "weak" promoters that bind poorly. These weak promoters usually permit a lower rate of transcription than the strong promoters, because the transcription machinery binds to them and initiates transcription less frequently. Other possible regulatory regions include enhancers, which can compensate for a weak promoter. Most regulatory regions are "upstream"—that is, before or toward the 5' end of the transcription initiation site. Eukaryotic promoter regions are much more complex and difficult to identify than prokaryotic promoters.

Many prokaryotic genes are organized into operons, or groups of genes whose products have related functions and which are transcribed as a unit. By contrast, eukaryotic genes are transcribed only one at a time, but may include long stretches of DNA called introns which are transcribed but never translated into protein (they are spliced out before translation). Splicing can also occur in prokaryotic genes, but is less common than in eukaryotes.

Chromosomes

The total complement of genes in an organism or cell is known as its genome, which may be stored on one or more chromosomes; the region of the chromosome at which a particular gene is located is called its locus. A chromosome consists of a single, very long DNA helix on which thousands of genes are encoded. Prokaryotes—bacteria and archaea—typically store their genomes on a single large, circular chromosome, sometimes supplemented by additional small circles of DNA called plasmids, which usually encode only a few genes and are easily transferable between individuals. For example, the genes for antibiotic resistance are usually encoded on bacterial plasmids and can be passed between individual cells, even those of different species, via horizontal gene transfer. Although some simple eukaryotes also possess plasmids with small numbers of genes, the majority of eukaryotic genes are stored on multiple linear chromosomes, which are packed within the nucleus in complex with storage proteins called histones. The manner in which DNA is stored on the histone, as well as chemical modifications of the histone itself, are regulatory mechanisms governing whether a particular region of DNA is accessible for gene expression. The ends of eukaryotic chromosomes are capped by long stretches of repetitive sequences called telomeres, which do not code for any gene product but are present to prevent degradation of coding and regulatory regions during DNA replication. The length of the telomeres tends to decrease each time the genome is replicated in preparation for cell division; the loss of telomeres has been proposed as an explanation for cellular senescence, or the loss of the ability to divide, and by extension for the aging process in organisms.

Whereas the chromosomes of prokaryotes are relatively gene-dense, those of eukaryotes often contain so-called "junk DNA", or regions of DNA that serve no obvious function. Simple single-celled eukaryotes have relatively small amounts of such DNA, whereas the genomes of complex multicellular organisms, including humans, contain an absolute majority of DNA without an identified function. However it now appears that, although protein-coding DNA makes up barely 2% of the human genome, about 80% of the bases in the genome may be being expressed, so the term "junk DNA" may be a misnomer.

Gene Expression

In all organisms, there are two major steps separating a protein-coding gene from its protein: First, the DNA on which the gene resides must be *transcribed* from DNA to messenger RNA (mRNA); and, second, it must be *translated* from mRNA to protein. RNA-coding genes must still go through the first step, but are not translated into protein. The process of producing a biologically functional molecule of either RNA or protein is called gene expression, and the resulting molecule itself is called a gene product.

Genetic Code

The genetic code is the set of rules by which a gene is translated into a functional protein. Each gene consists of a specific sequence of nucleotides encoded in a DNA (or sometimes RNA) strand; a correspondence between nucleotides, the basic building blocks of genetic material, and amino acids, the basic building blocks of proteins, must be established for genes to be successfully translated into functional proteins. Sets of three nucleotides, known as codons, each correspond to a specific amino acid or to a signal; three codons are known as "stop codons" and, instead of specifying a new amino acid, alert the translation machinery that the end of the gene has been reached. There are 64 possible codons (four possible nucleotides at each of three positions, hence 4^3 possible codons) and only 20 standard amino acids; hence the code is redundant and multiple codons can specify the same amino acid. The correspondence between codons and amino acids is nearly universal among all known living organisms.

Transcription

The process of genetic transcription produces a single-stranded RNA molecule known as messenger RNA, whose nucleotide sequence is complementary to the DNA from which it was transcribed. The DNA strand whose sequence matches that of the RNA is known as the coding strand and the strand from which the RNA was synthesized is the template strand. Transcription is performed by an enzyme called an RNA polymerase, which reads the template strand in the 3' to 5' direction and synthesizes the RNA from 5' to 3'. To initiate transcription, the polymerase first recognizes and binds a promoter region of the gene. Thus a major mechanism of gene regulation is the blocking or sequestering of the promoter region, either by tight binding by repressor molecules that physically block the polymerase, or by organizing the DNA so that the promoter region is not accessible.

In prokaryotes, transcription occurs in the cytoplasm; for very long transcripts, translation may begin at the 5' end of the RNA while the 3' end is still being transcribed. In eukaryotes, transcription necessarily occurs in the nucleus, where the cell's DNA is sequestered; the RNA molecule produced by the polymerase is known as the primary transcript and must undergo post-transcriptional modifications before being exported to the cytoplasm for translation. The splicing of introns present within the transcribed region is a modification unique to eukaryotes; alternative splicing mechanisms can result in mature transcripts from the same gene having different sequences and thus coding for different proteins. This is a major form of regulation in eukaryotic cells.

Translation

Translation is the process by which a mature mRNA molecule is used as a template for synthesizing a new protein. Translation is carried out by ribosomes, large complexes of RNA and protein responsible for carrying out the chemical reactions to add new amino acids to a growing polypeptide chain by the formation of peptide bonds. The genetic code is read three nucleotides at a time, in units called codons, via interactions with specialized RNA molecules called transfer RNA (tRNA). Each tRNA has three unpaired bases known as the anticodon that are complementary to the codon it reads; the tRNA is also covalently attached to the amino acid specified by the complementary codon. When the tRNA binds to its complementary codon in an mRNA strand, the ribosome ligates its amino acid cargo to the new polypeptide chain, which is synthesized from amino terminus to carboxyl terminus. During and after its synthesis, the new protein must fold to its active three-dimensional structure before it can carry out its cellular function.

DNA Replication and Inheritance

The growth, development, and reproduction of organisms relies on cell division, or the process by which a single cell divides into two usually identical daughter cells. This requires first making a duplicate copy of every gene in the genome in a process called DNA replication. The copies are made by specialized enzymes known as DNA polymerases, which "read" one strand of the double-helical DNA, known as the template strand, and synthesize a new complementary strand. Because the DNA double helix is held together by base pairing, the sequence of one strand completely specifies the sequence of its complement; hence only one strand needs to be read by the enzyme to produce a faithful copy. The process of DNA replication is semiconservative; that is, the copy of the genome inherited by each daughter cell contains one original and one newly synthesized strand of DNA.

After DNA replication is complete, the cell must physically separate the two copies of the genome and divide into two distinct membrane-bound cells. In prokaryotes - bacteria and archaea - this usually occurs via a relatively simple process called binary fission, in which each circular genome attaches to the cell membrane and is separated into the daughter cells as the membrane invaginates to split the cytoplasm into two membrane-bound portions. Binary fission is extremely fast compared to the rates of cell division in eukaryotes. Eukaryotic cell division is a more complex process known as the cell cycle; DNA replication occurs during a phase of this cycle known as S phase, whereas the process of segregating chromosomes and splitting the cytoplasm occurs during M phase. In many single-celled eukaryotes such as yeast, reproduction by budding is common, which results in asymmetrical portions of cytoplasm in the two daughter cells.

Molecular Inheritance

The duplication and transmission of genetic material from one generation of cells to the next is the basis for molecular inheritance, and the link between the classical and molecular pictures of genes. Organisms inherit the characteristics of their parents because the cells of the offspring contain copies of the genes in their parents' cells. In asexually reproducing organisms, the offspring will be a genetic copy or clone of the parent organism. In sexually reproducing organisms, a specialized form of cell division called meiosis produces cells called gametes or germ cells that are haploid, or contain only one copy of each gene. The gametes produced by females are called eggs or ova, and those produced by males are called sperm. Two gametes fuse to form a fertilized egg, a single cell that once again has a diploid number of genes—each with one copy from the mother and one copy from the father. During the process of meiotic cell division, an event called genetic recombination or *crossing-over* can sometimes occur, in which a length of DNA on one chromatid is swapped with a length of DNA on the corresponding sister chromatid. This has no effect if the alleles on the chromatids are the same, but results in reassortment of otherwise linked alleles if they are different. The Mendelian principle of independent assortment asserts that each of a parent's two genes for each trait will sort independently into gametes; which allele an organism inherits for one trait is unrelated to which allele it inherits for another trait. This is in fact only true for genes that do not reside on the same chromosome, or are located very far from one another on the same chromosome. The closer two genes lie on the same chromosome, the more closely they will be associated in gametes and the more often they will appear together; genes that are very close are essentially never separated because it is extremely unlikely that a crossover point will occur between them. This is known as genetic linkage.

History

Prior to Mendel's work, the dominant theory of heredity was one of blending inheritance, which proposes that the traits of the parents blend or mix in a smooth, continuous gradient in the offspring. Although Mendel's work was largely unrecognized after its first publication in 1866, it was rediscovered in 1900 by three European scientists, Hugo de Vries, Carl Correns, and Erich von Tschermak, who had reached similar conclusions from their own research. However, these scientists were not yet aware of the identity of the 'discrete units' on which genetic material resides.

The existence of genes was first suggested by Gregor Mendel (1822–1884), who, in the 1860s, studied inheritance in peaplants (*Pisum sativum*) and hypothesized a factor that conveys traits from parent to offspring. He spent over 10 years of his life on one experiment. Although he did not use the term *gene*, he explained his results in terms of inherited characteristics. Mendel was also the first to hypothesize independent assortment, the distinction between dominant and recessive traits, the distinction between a heterozygote and homozygote, and the difference between what would later be described as genotype (the genetic material of an organism) and phenotype (the visible traits of that organism). Mendel's concept was given a name by Hugo de Vries in 1889, who, at that time probably unaware of Mendel's work, in his book *Intracellular Pangenesis* coined the term "pangen" for "the smallest particle representing] one hereditary characteristic".

Darwin used the term Gemmule to describe a microscopic unit of inheritance, and what would later become known as Chromosomes had been observed separating out during cell division by Wilhelm Hofmeister as early as 1848. The idea that chromosomes are the carriers of inheritance was expressed in 1883 by Wilhelm Roux.

The modern conception of the gene originated with work by Gregor Mendel, a 19th-century Augustinian monk who systematically studied heredity in pea plants. Mendel's work was the first to illustrate particulate inheritance, or the theory that inherited traits are passed from one generation to the next in discrete units that interact in well-defined ways. Danish botanist Wilhelm Johannsen coined the word "gene" ("gen" in Danish and German) in 1909 to describe these fundamental physical and functional units of heredity, while the related word genetics was first used by William Bateson in 1905. The word was derived from Hugo de Vries' 1889 term *pangen* for the same concept, itself a derivative of the word *pangenesis* coined by Darwin (1868). The word pangenesis is made from the Greek words *pan* (a prefix meaning "whole", "encompassing") and *genesis* ("birth") or *genos* ("origin").

In the early 1900s, Mendel's work received renewed attention from scientists. In 1910, Thomas Hunt Morgan showed that genes reside on specific chromosomes. He later showed that genes occupy specific locations on the chromosome. With this knowledge, Morgan and his students began the first chromosomal map of the fruit fly *Drosophila*. In 1928, Frederick Griffith showed that genes could be transferred. In what is now known as Griffith's experiment, injections into a mouse of a deadly strain of bacteria that had been heat-killed transferred genetic information to a safe strain of the same bacteria, killing the mouse.

A series of subsequent discoveries led to the realization decades later that chromosomes within cells are the carriers of genetic material, and that they are made of DNA (deoxyribonucleic acid), a polymeric molecule found in all cells on which the 'discrete units' of Mendelian inheritance are encoded. In 1941, George Wells Beadle and Edward Lawrie Tatum showed that mutations in genes caused errors in specific steps in metabolic pathways. This showed that specific genes code for specific proteins, leading to the "one gene, one enzyme" hypothesis. Oswald Avery, Colin Munro MacLeod, and Maclyn McCarty showed in 1944 that DNA holds the gene's information. In 1953, James D. Watson and Francis Crick demonstrated the molecular structure of DNA. Together, these discoveries established the central dogma of molecular biology, which states that proteins are translated from RNA which is transcribed from DNA.

In 1972, Walter Fiers and his team at the Laboratory of Molecular Biology of the University of Ghent (Ghent, Belgium) were the first to determine the sequence of a gene: the gene for Bacteriophage MS2 coat protein. Richard J. Roberts and Phillip Sharp discovered in 1977 that genes can be split into segments. This led to the idea that one gene can make several proteins. Recently (as of 2003–2006), biological results let the notion of gene appear more slippery. In particular, genes do not seem to sit side by side on DNA like discrete beads. Instead, regions of the DNA producing distinct proteins may overlap, so that the idea emerges that "genes are one long continuum".

It was first hypothesized in 1986 by Walter Gilbert that neither DNA nor protein would be required in such a primitive system as that of a very early stage of the earth if RNA could perform as simply a catalyst and genetic information storage processor. The modern study of genetics at the level of DNA is known as molecular genetics and the synthesis of molecular genetics with traditional Darwinian evolution is known as the modern evolutionary synthesis.

Mendelian Inheritance and Classical Genetics

According to the theory of Mendelian inheritance, variations in phenotype—the observable physical and behavioral characteristics of an organism—are due to variations in genotype, or the organism's particular set of genes, each of which specifies a particular trait. Different forms of a gene, which may give rise to different phenotypes, are known as alleles. Organisms such as the pea plants Mendel worked on, along with many plants and animals, have two alleles for each trait, one inherited from each parent. Alleles may be dominant or recessive; dominant alleles give rise to their corresponding phenotypes when paired with any other allele for the same trait, whereas recessive alleles give rise to their corresponding phenotype only when paired with another copy of the same allele.

For example, if the allele specifying tall stems in pea plants is dominant over the allele specifying short stems, then pea plants that inherit one tall allele from one parent and one short allele from the other parent will also have tall stems. Mendel's work found that alleles assort independently in the production of gametes, or germ cells, ensuring variation in the next generation.

Mutation

DNA replication is for the most part extremely accurate, with an error rate per site of around 10^{-6} to 10^{-10} in eukaryotes. Rare, spontaneous alterations in the base sequence of a particular gene arise from a number of sources, such as errors in DNA replication and the aftermath of DNA damage. These errors are called mutations. The cell contains many DNA repair mechanisms for preventing mutations and maintaining the integrity of the genome; however, in some cases—such as breaks in both DNA strands of a chromosome—repairing the physical damage to the molecule is a higher priority than producing an exact copy. Due to the degeneracy of the genetic code, some mutations in protein-coding genes are *silent*, or produce no change in the amino acid sequence of the protein for which they code; for example, the codons UCU and UUC both code for serine, so the U↔C mutation has no effect on the protein. Mutations that do have phenotypic effects are most often neutral or deleterious to the organism, but sometimes they confer benefits to the organism's fitness.

Mutations propagated to the next generation lead to variations within a species' population. Variants of a single gene are known as alleles, and differences in alleles may give rise to differences in traits. Although it is rare for the variants in a single gene to have clearly distinguishable phenotypic effects, certain well-defined traits are in fact controlled by single genetic loci. A gene's most common allele is called the wild type allele, and rare alleles are called mutants. However, this does not imply that the wild-type allele is the ancestor from which the mutants are descended.

Genome

Chromosomal Organization

The total complement of genes in an organism or cell is known as its genome. In prokaryotes, the vast majority of genes are located on a single chromosome of circular DNA, while eukaryotes usually possess multiple individual linear DNA helices packed into dense DNA-protein complexes called chromosomes. Genes that appear together on one chromosome of one species may appear on separate chromosomes in another species. Many species carry more than one copy of their genome within each of their somatic cells. Cells or organisms with only one copy of each chromosome are called haploid; those with two copies are called diploid; and those with more than two copies are called polyploid. The copies of genes on the chromosomes are not necessarily identical. In sexually reproducing organisms, one copy is normally inherited from each parent.

Number of Genes

Early estimates of the number of human genes that used expressed sequence tag data put it at 50 000–100 000. Following the sequencing of the human genome and other genomes, it has been found that rather few genes (~20 000 in human, mouse and fly, ~13 000 in roundworm, >46 000 in rice) encode all the proteins in an organism. These protein-coding sequences make up 1–2% of the human genome. A large part of the genome is transcribed however, to introns, retrotransposons and seemingly a large array of noncoding RNAs. Total number of proteins (the Earth's proteome) is estimated to be 5 million sequences.

Genetic and Genomic Nomenclature

Gene nomenclature has been established by the HUGO Gene Nomenclature Committee (HGNC) for each known human gene in the form of an approved gene name and symbol (short-form abbreviation). All approved symbols are stored in the HGNC Database. Each symbol is unique and each gene is only given one approved gene symbol. This also facilitates electronic data retrieval from publications. In preference each symbol maintains parallel construction in different members of a gene family and can be used in other species, especially the mouse.

Evolutionary Concept of a Gene

George C. Williams first explicitly advocated the gene-centric view of evolution in his 1966 book *Adaptation and Natural Selection*. He proposed an evolutionary concept of gene to be used when we are talking about natural selection favouring some genes. The definition is: "that which segregates and recombines with appreciable frequency." According to this definition, even an asexual genome could be considered a gene, insofar that it have an appreciable permanency through many generations.

The difference is: the molecular gene *transcribes* as a unit, and the evolutionary gene *inherits* as a unit.

Richard Dawkins' books *The Selfish Gene* (1976) and *The Extended*

Phenotype (1982) defended the idea that the gene is the only replicator in living systems. This means that only genes transmit their structure largely intact and are potentially immortal in the form of copies. So, genes should be the unit of selection. In *The Selfish Gene* Dawkins attempts to redefine the word 'gene' to mean "an inheritable unit" instead of the generally accepted definition of "a section of DNA coding for a particular protein".

In *River Out of Eden*, Dawkins further refined the idea of gene-centric selection by describing life as a river of compatible genes flowing through geological time. Scoop up a bucket of genes from the river of genes, and we have an organism serving as temporary bodies or survival machines. A river of genes may fork into two branches representing two non-interbreeding species as a result of geographical separation.

Gene Targeting and Implications

Gene targeting is commonly referred to techniques for altering or disrupting mouse genes and provides the mouse models for studying the roles of individual genes in embryonic development, human disorders, aging and diseases. The mouse models, where one or more of its genes are deactivated or made inoperable, are called knockout mice. Since the first reports in which homologous recombination in embryonic stem cells was used to generate gene-targeted mice, gene targeting has proven to be a powerful means of precisely manipulating the mammalian genome, producing at least ten thousand mutant mouse strains and it is now possible to introduce mutations that can be activated at specific time points, or in specific cells or organs, both during development and in the adult animal.

Gene targeting strategies have been expanded to all kinds of modifications, including point mutations, isoform deletions, mutant allele correction, large pieces of chromosomal DNA insertion and deletion, tissue specific disruption combined with spatial and temporal regulation and so on. It is predicted that the ability to generate mouse models with predictable phenotypes will have a major impact on studies of all phases of development, immunology, neurobiology, oncology, physiology, metabolism, and human diseases. Gene targeting is also in theory applicable to species from which totipotent embryonic stem cells can be established, and therefore may offer a potential to the improvement of domestic animals and plants.

Changing Concept

The concept of the gene has changed considerably. From the original definition of a "unit of inheritance", the term evolved to mean a DNA-based unit that can exert its effects on the organism through RNA or protein products. It was also previously believed that one gene makes one protein; this concept was overthrown by the discovery of alternative splicing and trans-splicing.

The definition of a gene is still changing. The first cases of RNA-based inheritance have been discovered in mammals. Evidence is also accumulating that the control regions of a gene do not necessarily have to be close to the coding sequence on the linear molecule or even on the same chromosome. Spilianakis and colleagues discovered that the promoter region of the interferon-gamma gene on chromosome 10 and the regulatory regions of the T(H)2 cytokine locus on chromosome 11 come into close proximity in the nucleus possibly to be jointly regulated. The concept that genes are clearly delimited is also being eroded. There is evidence for fused proteins stemming from two adjacent genes that can produce two separate protein products.

While it is not clear whether these fusion proteins are functional, the phenomenon is more frequent than previously thought. Even more ground-breaking than the discovery of fused genes is the observation that some proteins can be composed of exons from far away regions and even different chromosomes. This new data has led to an updated, and probably tentative, definition of a gene as "a union of genomic sequences encoding a coherent set of potentially overlapping functional products." This new definition categorizes genes by functional products, whether they be proteins or RNA, rather than specific DNA loci; all regulatory elements of DNA are therefore classified as *gene-associated* regions.

A single gene coding for a particular protein corresponds to a sequence of nucleotides along one or more regions of a molecule of DNA. The DNA sequence is collinear with the protein sequence. In species for which the genetic materiais double-stranded DNA, genes may appear on either strand. Bacterial genes are continuous regions of DNA. Therefore the functional unit of genetic sequence information from a bacterium is a string of $3N$ nucleotides encoding a string of N amino acids, or a string of N nucleotides encoding a structural RNA molecule. Such a string, equipped with annotations, would form a typical entry in one of the genetic sequence archives.

Proteomes

The proteome project is a large-scale programme dealing in an integral way with patterns of expression of proteins in biological systems, in ways that complement and extend genome projects.

What kinds of data would we like to measure, and what mature experimental techniques exist to determine them? The basic goal is a spatio-temporal description of the deployment of proteins in the organism. The rates of synthesis of different proteins vary among different tissues and different cell types and states of activity. Methods are available for analysis of transcription patterns of genes. However, because proteins 'turn over' at different rates, it is also necessary to measure proteins directly. High-resolution two-dimension-apolyacylamide gel electrophoresis (2D PAGE) shows the pattern of protein content in a sample. Mass-spectroscopic techniques identify the proteins into which the sample has been separated.

In principle, a database of amino acid sequences of proteins is inherent in the database of nucleotide sequences of DNA, by virtue of the genetic code. Indeed, new protein sequence data are now being determined almost exclusively by translation of DNA sequences, rather than by direct sequencing of proteins. (Historically the chemical problem of determining amino acid sequences of proteins directly was solved before the genetic code was established and before methods for determination of nucleotide sequences of DNA were developed. F. Sanger's sequencing of insulin in 1955 first proved that proteins had definite amino acid sequences, a proposition that until then was hypothetical.) Should any distinction be made between amino acid sequences determined directly from proteins and those determined by translation from DNA? First, we must assume that it is possible correctly to identify within the DNA data stream the regions that encode proteins.

The pattern-recognition programs that address this question are subject to three types of errors: a genuine protein sequence mabe missed entirely, or an incomplete protein may be reported, or a gene may be incorrectly spliced. Several variations on the theme add to the complexity: Genes for different proteins may overlap, or genes may be assembled from exons in different ways in different tissues. Conversely, some genetic sequences that appear to code for proteins may in fact be defective or not expressed. *A protein inferred* from a genome sequence is a hypothetical object until an experiment verifies its existence.

Second, in many cases the expression of a gene produces a molecule that must be modified within a cell, to make a *mature* protein that differs significantly frothe one suggested by translation of the gene sequence. In many cases the missing details of post-translational modifications—the molecular analogues of body piercing—are quite important. Post-translational modifications include addition of ligands (for instance the covalently-bound haem group of cytochrome *c*), glycosylation, methylation, excision of peptides, and many others. Patterns of disulphide bridges—primary chemical bonds between cysteine residues—cannot be deduced from the amino acid sequence. In some cases, mRNA is edited before translation, creating changes in amino acid sequences that are not inferrable from the genes.

Eavesdropping on the Transmission of Genetic Information

How hereditary information is stored, passed on, and implemented is perhaps *the* fundamental problem of biology. Three types of maps have been essential :

1. Linkage maps of genes
2. Banding patterns of chromosomes
3. DNA sequences.

These represent three very different types of data. Genes, as discovered by Mendel, were entirely abstract entities. Chromosomes are physical objects, the banding patterns their visible landmarks. Only with DNA sequences are we dealing directly with stored hereditary information in its physical form.

It was the very great achievement of the last century of biology to forge connections between these three types of data. The first steps—and giant strides thewere indeed—proved that, for any chromosome, the maps are one-dimensional arrays, and indeed that they are collinear. Any schoolchild now knows that genes are strung out along chromosomes, and that each gene corresponds to a DNA sequence. But the proofs of these statements earned a large number of Nobel prizes.

Splitting a long molecule of DNA—for example, the DNA in an entire chromosome— into fragments of convenient size for cloning and sequencing requires additional maps to report the order of the fragments, so that the entire sequence can be reconstructed from the sequences of the fragments. A restriction endonuclease is an enzyme that cuts DNA at a specific sequence, usually about 6 blong. Cutting

DNA with several restriction enzymes with different specificities produces sets of overlapping fragments. From the sizes of the fragments it is possible to construct a restriction map, stating the order and distance between the restriction enzyme cleavage sites. A mutation in one of these cleavage sites will change the sizes of the fragments produced by the corresponding enzyme, allowing the mutation to be located in the map.

Restriction enzymes can produce fairly large pieces of DNA. Cutting the DNinto smaller pieces, which are cloned and ordered by sequence overlaps (like the example using text on pages 19–20) produces a finer dissection of the DNA called a contig map.

In the past, the connections between chromosomes, genes and DNA sequences have been essential for identifying the molecular deficits underlying inheritediseases, such as Huntington disease or cystic fibrosis. Sequencing of the human genome has changed the situation radically.

Given a disease attributable to a defective protein:

- If we know the protein involved, we can pursue rational approaches to therapy.
- If we know the gene involved, we can devise tests to identify sufferers or carriers.

In many cases, knowledge of the chromosomal location of the gene is unnecessary for either therapy or detection; it is required only for identifying the gene, providing a bridge between the patterns of inheritance and the DNA sequence. (This is not true of diseases arising from chromosome abnormalities.)

For instance, in the case of sickle-cell anaemia, we know the protein involved. The disease arises from a single point mutation in haemoglobin. We can proceed directly to drug design. We need the DNA sequence only for genetic testing and counselling. In contrast, if we know neither the protein nor the gene, we must somehow go from the phenotype back to the gene, a process called positional cloning or reverse genetics. Positional cloning used to involve a kind of 'Tinker to Evers to Chance' cascade from the gene map to the chromosome map to the DNA sequence.

Patterns of inheritance identify the type of genetic defect responsible for a condition. They show, for example, that Huntington disease and cystic fibrosis are caused by single genes. To find the gene associated with cystic fibrosis it was necessary to begin with the gene map, using linkage patterns of heredity in affected families to localize the affected gene to a particular region of a particular chromosome. Knowing the general region of the chromosome, it was then possible to search the DNA of that region to identify candidate genes, and finally to pinpoint the particular gene responsible and sequence it.

In contrast, many diseases do not show simple inheritance, or, even if only a single gene is involved, heredity creates only a predisposition, the clinical consequences of which depend on environmental factors. The full human genome sequence, and measurements of expression patterns, will be essential to identify the genetic components of these more complex cases.

Mappings between the Maps

A gene linkage map can be calibrated to chromosome banding patterns through observation of individuals with deletions or translocations of parts of chromosomes.

The genes responsible for phenotypic changes associated with a deletion must lie within the deletion. Translocations are correlated with altered patterns of linkage and recombination.

There have been several approaches to coordinating chromosome banding patterns with individual DNA sequences of genes:

- In Fluorescent *In Situ* Hybridization (FISH) a probe sequence is labelled with fluorescent dye. The probe is hybridized with the chromosomes, and the chromosomal location where the probe is bound shows up directly in a photograph. Typical resolution is ~105 base pairs, but specialized new techniques can achieve high resolution, down to 1 kbp. Simultaneous FISH with two probes can detect linkage and even estimate genetic distances. This is important in species for which the generation time is long enough to make standard genetic approaches inconvenient. FISH can also detect chromosomal abnormalities.
- Somatic cell hybrids are rodent cells containing few, one, or even partial human chromosomes. (Chromosome fragments are produced by irradiating the human cells prior to fusion. Such lines are called radiation hybrids.) Hybridization of a probe sequence with a panel of somatic cell hybrids, detected by fluorescence, can identify which chromosome contains the

probe. This approach has been superseded by use of clones of yeast or bacteria or phage containing fragments of human DNA in artificial chromosomes (YACs and BACs and PACs).

High-resolution Maps

Formerly, genes were the only visible portions of genomes. Now, markers are nlonger limited to genes with phenotypically observable effects, which are anyway too sparse for an adequately high-resolution map of the human genome. Now that we can interrogate DNA sequences directly, any features of DNA that vary among individuals can serve as markers, including:

Variable number tandem repeats (VNTRs), also called minisatellites. VNTRs contain regions 10–100 base pairs long, repeated a variable number of times— same sequence, different number of repeats. In any individual, VNTRs based on the same repeat motif may appear only once in the genome; or several times, with different lengths on different chromosomes. The distribution of the sizes of the repeats is the marker. Inheritance of VNTRs can be followed in a family and mapped to a disease phenotype like any other trait. VNTRs were the first genetic sequence data used for personal identification— genetic fingerprints— in paternity and in criminal cases.

Formerly, VNTRs were observed by producing restriction fragment length polymorphisms (RFLPs) from them. VNTRs are generally flanked by recognition sites for the same restriction enzyme, which will neatly excise them. The results can be spread out on a gel, and detected by Southern blotting. Note the distinction: VNTRs are characteristics of genome sequences; RFLPs are artificial mixtures of short stretches of DNA created in the laboratory in order to identify VNTRs.

It is much easier and more efficient to measure the sizes of VNTRs by amplifying them with PCR, and this method has replaced the use of restriction enzymes. Short tandem repeat polymorphisms (STRPs), also called microsatellites. STRPs are regions of only 2–5 bp but repeated many times; typically 10–30 consecutive copies. They have several advantages as markers over VNTRs, one of which is a more even distribution over the human genome. There is no reason why these markers need lie within expressed genes, and usually they do not. (The CAG repeats in the gene for huntingtin and certain other disease genes are exceptions.)

Panels of microsatellite markers greatly simplify the identification of genes. It is interesting to compare a recent project to identify a disease gene now that the human genome sequence is available, with such classic studies as the identification of the gene for cystic fibrosis.

Additional mapping techniques deal more directly with the DNA sequences, and can short-circuit the process of gene identification:

A contig, or contiguous clone map, is a series of overlapping DNA clones of known order along a chromosome from an organism of interest—for instance, human—stored in yeast or bacterial cells as YACs (Yeast Artificial Chromosomes) or BACs (Bacterial Artificial Chromosomes). A contig map can produce a very fine mapping of a genome. In a YAC, human DNA is stably integrated into a small extra chromosome in a yeast cell. A YAC can contain up to 106 base pairs.

In principle, the entire human genome could be represented in 10 000 YAC clones. In a BAC, human DNA is inserted into a plasmid in an *E. coli* cell. (A plasmid is a small piece of double-stranded DNA found in addition to the main genome, usually but not always circular.) A BAC can carry about 250 000 bp. Despite their smaller capacities, BACs are preferred to YACs because of their greater stability and ease of handling. A sequence tagged site (STS) is a short, sequenced region of DNA, typically 200–600 bp long, that appears in a unique location in the genome. It need not be polymorphic. An STS can be mapped into the genome by using PCR to test for the presence of the sequence in the cells containing a contig map.

One type of STS arises from an expressed sequence tag (EST), a piece of cDNA (complementary DNA; that is, a DNA sequence derived from the messenger RNA of an expressed gene). The sequence contains only the exons of the gene, spliced together to form the sequence that encodes the protein. cDNA sequences can be mapped to chromosomes using FISH, or located within contig maps.

How do contig maps and sequence tagged sites facilitate identifying genes? If you are working with an organism for which the full-genome sequence is not known, but for which full contig maps are available for all chromosomes, you would identify STS markers tightly linked to your gene, then locate these markers in the contig maps.

Picking out Genes in Genomes

Computer programs for genome analysis identify open reading frames or ORFs. An ORF is a region of DNA sequence that begins with an initiation codon (ATG) and ends with a stop codon. An ORF is a potential protein-coding region. Approaches to identifying protein-coding regions choose from or combine two possible approaches:

(1) Detection of regions similar to known coding regions from other organisms. These regions may encode amino acid sequences similar to known proteins, or may be similar to Expressed Sequence Tags (ESTs). Because ESTs are derived from messenger RNA, they correspond to genes known to be transcribed. It is necessary to sequence only a few hundred initial bases of cDNA to give enough information to identify a gene: characterization of genes by ESTs is like indexing poems or songs by their first lines.

(2) *Ab initio* methods, that seek to identify genes from the properties of the DNA sequences themselves. Computer-assisted annotation of genomes is more complete and accurate for bacteria than for eukaryotes. Bacterial genes are relatively easy to identify because they are contiguous—they lack the introns characteristic of eukaryotic genomes—and the intergene spaces are small. In higher organisms, identifying genes is harder. Identification of exons is one problem, assembling them is another. Alternative splicing patterns present a particular difficulty.

A framework for *ab initio* gene identification in eukaryotic genomes includes the following features:

The initial (5_) exon starts with a transcription start point, preceded by a corpromotor site such as the TATA box typically ~30 bp upstream. It is free of in-frame stop codons, and ends immediately before a GT splice signal. (Occasionally a noncoding exon precedes the exon that contains the initiator codon.)

Internal exons, like initial exons, are free of in-frame stop codons. They begin immediately after an AG splice signal and end immediately before a GT splice signal.

The final (3_) exon starts immediately after an AG splice signal and ends with a stop codon, followed by a polyadenylation signal sequence. (Occasionally a noncoding exon follows the exon that contains the stop codon.)

All coding regions have nonrandom sequence characteristics, based partly on codon usage preferences. Empirically, it is found that statistics of hexanucleotides perform best in distinguishing coding from noncoding regions. Starting from a set of known genes from an organism as a training set, pattern recognition programs can be tuned to particular genomes.

Accurate gene detection is a crucial component of genome sequence analysis. This problem is an important focus of current research.

Genomes of Prokaryotes

Most prokaryotic cells contain their genetic material in the form of a large single circular piece of double-stranded DNA, usually less than 5 Mbp long. In addition they may contain plasmids. The protein-coding regions of bacterial genomes do not contain introns. In many prokaryotic genomes the protein-coding regions are partially organized into *operons*—tandem genes transcribed into a single messenger RNA molecule under common transcriptional control. In bacteria, the genes of many operons code for proteins with related functions. For instance, successive genes in the trp operon of *E. coli* code for proteins that catalyse successive steps in the biosynthesis of tryptophan. In archaea, a metabolic relationship between genes in operons is less frequently observed.

The Genome of the Bacterium Escherichia Coli

E. coli, strain K-12, has long been the workhorse of molecular biology. The genome of strain MG1655, published in 1997 by the group of F. Blattner at the University of Wisconsin, contains 4 639 221 bp in a single circular DNA molecule, with no plastids. Approximately 89% of the sequence codes for proteins or structural RNAs. An inventory reveals:

- 4284 protein-coding genes
- 122 structural RNA genes
- noncoding repeat sequences
- regulatory elements
- transcription/translation guides
- transposases
- prophage remnants
- insertion sequence elements
- patches of unusual composition, likely to be foreign elements introduced by horizontal transfer.

Analysis of the genome sequence required identification and annotation of protein-coding genes and other functional regions. Many *E. coli* proteins were known before the sequencing was complete, from the many years of intensive investigation: 1853 proteins had been described before publication of the genome sequence. Other genes could be assigned functions from identification of homologues by searching in sequence databanks. The narrower the range of specificity of the function of the homologues, the more precise could be the assignment. Currently, over 80% of proteins can be assigned at least a general function. Other regions of the genome are recognized as regulatory sites, or mobilgenetic elements, also on the basis of similarity to homologous sequences known in other organisms.

The distribution of protein-coding genes over the genome of *E. coli* does not seem to follow any simple rules, either along the DNA or on different strands. Indeed, comparison of strains suggests that genes are mobile.

The *E. coli* genome is relatively gene dense. Genes coding for proteins or structural RNAs occupy ~89% of the sequence. The average size of an open reading frame is 317 amino acids. If the genes were evenly distributed, the average intergenic region would be 130 bp; the observed average distance between genes is 118 bp. However, the sizes of intergenic regions vary considerably. Some intergenic regions are large. These contain sites of regulatory function, and repeated sequences. The longest intergenic region, 1730 bp, contains noncoding repeat sequences.

Approximately three-quarters of the transcribed units contain only one gene; the rest contain several consecutive genes, or operons. It is estimated that the *E. coli* genome contains 630–700 operons. Operons vary in size, although few contain more than five genes. Genes within operons tend to have related functions.

In some cases, the same DNA sequence encodes parts of more than one polypeptide chain. One gene codes for both the and _ subunits of DNA polymerase III.

Translation of the entire gene forms the subunit. The _ subunit corresponds approximately to the N-terminal two-thirds of the subunit. A frameshift on the ribosome at this point leads to chain termination 50% of the time, causing a 1:1 ratio of expressed and _ subunits. There do not appear to be any overlapping genes in which different reading frames both code for expressed proteins. In other cases, the same polypeptide chain appears in more than one enzyme. A protein that functions on its own as lipoate dehydrogenase is also an essential subunit of pyruvate dehydrogenase, 2-oxoglutarate dehydrogenase and the glycine cleavage complex.

Having the complete genome, we can examine the protein repertoire of *E. coli*. The largest class of proteins are the enzymes—appoximately 30% of the total genes. Many enzymatic functions are shared by more than one protein. Some of these sets of functionally similar enzymes are very closely related, and appear to have arisen by duplication, either in *E. coli* itself or in an ancestor. Other sets of functionally similar enzymes have very dissimilar sequences, and differ in specificity, regulation, or intracellular location.

Several features of *E. coli's* generous endowment of enzymes give it an extensive and versatile metabolic competence, which allow it to grow and compete under varying conditions:

- It can synthesize all components of proteins and nucleic acids (amino acids and nucleotides), and cofactors.
- It has metabolic flexibility: Both aerobic and anaerobic growth are possible, utilizing different pathways of energy capture. It can grow on many different carbon sources. Not all metabolic pathways are active at any time, the versatility allows response to changes in conditions.
- Even for specific metabolic reactions there are many cases of multiple enzymes. These provide redundancy, and contribute to an ability to tune metabolism to varying conditions.
- However, *E. coli* does not possess a complete range of enzymatic capacity. It cannot fix CO2 or N2.

We have described here some of the *static* features of the *E. coli* genome and itprotein repertoire. Current research is moving into dynamic aspects, to investigations of protein expression patterns in time and space.

The Genome of the Archaeon Methanococcus Jannaschii

S. Luria once suggested that to determine common features of all life one should not try to survey everything, but rather identify the organism most different from us and see what we have in common with it. The assumption was that the way to do this would be to find an organism adapted to the most different environment. Deep-sea exploration has revealed environments as far from the familiar as those portrayed in science fiction. Hydrothermal vents are underwater volcanoes emitting hot lava and gases through cracks in the ocean floor. They create niches for communities of living things disconnected from the surface, depending on the minerals exuded from the vent for their sources of inorganic nutrients.

They support living communities of micro-organisms that are the only known forms of life not dependent on sunlight, directly or indirectly, as their energy source.

The micro-organism *Methanococcus jannaschii* was collected from a hydrothermal vent 2600 m deep off the coast of Baja California, Mexico, in 1983. It is a thermophilic organism, surviving at temperatures from 48–94 _C, with aoptimum at 85 _C. It is a strict anaerobe, capable of self-reproduction from inorganic components. Its overall metabolic equation is to synthesize methane from H2 and CO2.

M. jannaschii belongs to the archaea, one of the three major divisions of life along with the bacteria and eukarya. The archaea comprise groups of prokaryotes, including organisms adapted to extreme environmental conditions such as high temperature and pressure, or high salt concentration. The genome of *M. jannaschii* was sequenced in 1996 by The Institute for Genomic Research (TIGR). It was the first archaeal genome sequenced. It contains a large chromosome containing a circular double-stranded DNA molecule 1664 976 bp long, and two extrachromosomal elements of 58 407 and 16 550 bp respectively.

There are 1743 predicted coding regions, of which 1682 are on the chromosome, and 44 and 12 are on the large and small extrachromosomal elements respectively. Some RNA genes contain introns. As in other prokaryoti genomes there is little noncoding DNA. *M. jannaschii* would appear to satisfy Luria's goal of finding our most distant extant relative. Comparison of its genome sequence with others shows that it is distantly related to other forms of life. Only 38% of the open reading frames could be assigned a function, on the basis of homology to proteins known from other organisms. However, to everyone's great surprise, archaea are in some ways more closely related to eukaryotes than to bacteria! They are a complex mixture. Iarchaea, proteins involved in transcription, translation and regulation are more similar to those of eukaryotes. Archaeal proteins involved in metabolism are more similar to those of bacteria.

The Genome of One of the Simplest Organisms

Mycoplasma Genitalium

Mycoplasma genitalium is an infectious bacterium, the cause of nongonococcal uretheitis. Its genome was sequenced in 1995 by a collaboration of groups at TIGR, The Johns Hopkins University and The University of North Carolina. The genome is a single DNA molecule containing 580 070 base pairs. This is the smallest cellular genome yet sequenced. So far, *M. genitalium* is the closest we have to a minimal organism, the smallest capable of independent life. (Viruses, in contrast, require the cellular machinery of their hosts.)

The genome is dense in coding regions. 468 genes have been identified as expressed proteins. Some regions of the sequence are gene rich, others gene poor, but overall 85% of the sequence is coding. The average length of a coding region is 1040 bp. As in other bacterial genomes, the coding regions do not contain introns. Further compression of the genome is achieved by overlapping genes. It appears that many of these have arisen through loss of stop codons. The gene repertoire of *M. genitalium* includes some that encode proteins essential for independent reproduction, such as those involved in DNA replication, transcription, and translation, plus ribosomal and transfer RNAs. Other genes are specific for the infectious activity, including adhesins that mediate binding to infected cells, other molecules for defence against the host's immune system, and many transport proteins. As adaptation to the parasitic lifestyle of the organism, there has been widespread loss of metabolic enzymes, including those responsible for amino acid biosynthesis—indeed, one amino acid is absent from all *M. genitalium* proteins.

Genomes of Eukaryotes

It is rare in science to encounter a completely new world containing phenomena entirely unsuspected. The complexity of the eukaryotic genome is such a world In eukaryotic cells, the majority of DNA is in the nucleus, separated into bundles of nucleoprotein, the chromosomes. Each chromosome contains a single double-stranded DNA molecule. Smaller amounts of DNA appear in organelles—mitochondria and chloroplasts. The organelles originated as intracellular parasites. Organelle genomes usually have the form of circular doublestranded DNA, but are sometimes linear and sometimes appear as multiple circles. The genetic code by which organelle genes are translated differs from that of nuclear genes. Nuclear genomes of different species vary widely in size. The correlation between genome size and complexity of the organism is very rough. It certainly does not support any preconception that humans stand on a pinnacle. In many cases differences in genome size reflect different amounts of simple repetitive sequences.

In addition to variation in DNA content, eukaryotic species vary in the number of chromosomes and distribution of genes among them. Some differences in the distribution of genes among chromosomes involve translocations, or chromosome fragmentations or joinings. For instance, humans have 23 pairs of chromosomes; chimpanzees have 24. Human chromosome 2 is equivalent to a fusion of chimpanzee chromosomes 12 and 13. The difficulty of chromosome pairing during mitosis in a zygote after such an event can contribute to the reproductive isolation required for species separation.

Other differences in chromosome complement reflect duplication or hybridization events. The wheat first used in agriculture, in the Middle East at least 10 000–15 000 years ago, is a diploid called *einkorn* (*Triticum monococcum*), containing 14 pairs of chromosomes. *Emmer* wheat (*T. dicoccum*), also cultivated since palaeolithic times, and *durum* wheat (*T. turgidum*), are merged hybrids of relatives of einkorn with other wild grasses, to form tetraploid species. Additional hybridizations, to different wild wheats, gave hexaploid forms, including *spelt* (*T. spelta*), and modern common wheat *T. aestivum*. *Triticale*, a robust crop developed in modern agriculture and currently used primarily for animal feed, is an artificial genus arising from crossing durum wheat (*Triticum turgidum*) and rye (*Secale cereale*). Most triticale varieties are hexaploids.

All these species are still cultivated—some to only minor extents—and have their individual uses in cooking. Spelt, or *farro* in Italian, is the basis of a well-known soup; pasta is made from durum wheat; and bread from *T. aestivum*. Even within single chromosomes, gene families are common in eukaryotes. Some family members are paralogues—related genes that have diverged to provide separate functions in the same species. (Orthologues, in contrast, are homologues that perform the same function in different species. For instance, human and globin are paralogues, and human and horse myoglobin are orthologues.) Other related sequences may be pseudogenes, which may have arisen by duplication, or by retrotransposition from messenger RNA, followed by the accumulation of mutations to the point of loss of function. The human globin gene cluster is a good example.

The Genome of Saccharomyces Cerevisiae (Baker's Yeast)

Yeast is one of the simplest known eukaryotic organisms. Its cells, like our own, contain a nucleus and other specialized intracellular compartments. The sequencing of its genome, by an unusually effective international consortium involving ~100 laboratories, was completed in 1992. The yeast genome contains 12 057 500 bp of nuclear DNA, distributed over 16 chromosomes. The chromosomes range in size over an order of magnitude, from the 1352 kbp chromosome IV to the 230 kbp chromosome I.

The yeast genome contains 5885 predicted protein-coding genes, ~140 genes for ribosomal RNAs, 40 genes for small nuclear RNAs, and 275 transfer RNA genes. In two respects, the yeast genome is denser in coding regions than the known genomes of the more complex eukaryotes *Caenorhabditis elegans*, *Drosophila melanogaster*, and human: (1) Introns are relatively rare, and relatively small. Only 231 genes in yeast contain introns. (2) There are fewer repeat sequences compared with more complex eukaryotes.

A duplication of the entire yeast genome appears to have occurred ~150 million years ago. This was followed by translocations of pieces of the duplicated DNA and loss of one of the copies of most (~92%) of the genes. Of the 5885 potential protein-coding genes, 3408 correspond to known proteins. About 1000 more contain some similarity to known proteins in other species. Another ~800 are similar to ORFs in other genomes that correspond to unknown proteins. Many of these homologues appear in prokaryotes. Only approximately one-third of yeast proteins have identifiable homologues in the human genome.

In taking censuses of genes, it has been useful to classify their functions into broad categories.

- Metabolism
- Energy
- Cell growth, cell division and DNA synthesis
- Transcription
- Protein synthesis
- Protein destination

- Transport facilitation
- Cellular transport and transport mechanisms
- Cellular biogenesis
- Cellular communication/signal transduction
- Cell rescue, defence, cell death and ageing
- Ionic homeostasis
- Cellular organization
- Transposable elements, viral and plasmid proteins
- Unclassified.

Yeast is a testbed for development of methods to assign functions to gene products. The search for homologues has been exhaustive and continues. Collections of mutants exist containing a knockout of every gene. (A unique sequence 'bar code' introduced into each mutant facilitates identification of the ones that grow under selected conditions.) A surprisingly-large fraction of these single-gene knockout strains are viable—80%. Cellular localization and expression patterns are being investigated. Several types of measurements, including those based on activation of transcription by pairs of proteins that can form dimers—the two-hybrid method, are producing catalogues of interprotein interactions.

The Genome of Caenorhabditis Elegans

The nematode worm *Caenorhabditis elegans* entered biological research in the 1960s, at the express invitation of S. Brenner. He recognized its potential as a sufficiently complex organism to be interesting but simple enough to permit complete analysis, at the cellular level, of its development and neural circuitry. The *C. elegans* genome, completed in 1998, provided the first full DNA sequence of a multicellular organism. The *C. elegans* genome contains ~97 Mbp of DNA distributed on paired chromosomes I, II, III, IV, V and X. There is no Y chromosome. Different genders in *C. elegans* appear in the XX genotype, a selffertilizing hermaphrodite; and the XO genotype, a male.

Several kinds of RNA genes have been identified. The *C. elegans* genome contains 659 genes for tRNA, almost half of them (44%) on the X chromosome. Spliceosomal RNAs appear in dispersed copies, often identical. (Spliceosomes are the organelles that convert pre-mRNA transcripts to mature mRNA by excising introns.) Ribosomal

RNAs appear in a long tandem array at the end of chromosome I. 5S RNAs appear in a tandem array on chromosome V. Some RNA genes appear in introns of protein-coding genes. The *C. elegans* genome contains many repeat sequences. Approximately 2.6% of the genome consists of tandem repeats. Approximately 3.6% of the genome contains inverted repeats; these appear preferentially within introns, rather than between genes. Repeats of the hexamer sequence TTAGGC appear in many places. There are also simple duplications, involving hundreds to tens of thousands of kbp.

The Genome of Drosophila Melanogaster

Drosophila melanogaster, the fruit fly, has been the subject of detailed studies of genetics and development for a century. Its genome sequence, the product of a collaboration between Celera Genomics and the Berkeley Drosophila Genome Project, was announced in 1999.

The chromosomes of *D. melanogaster* are nucleoprotein complexes. Approximately one-third of the genome is contained in heterochromatin, highly coiled and compact (and therefore densely staining) regions flanking the centromeres. The other two-thirds is euchromatin, a relatively uncoiled, less-compact form. Most of the active genes are in the euchromatin. The heterochromatin in *D. melanogaster* contains many tandem repeats of the sequence AATAACATAG, and relatively few genes. The total chromosomal DNA of *Drosophila melanogaster* contains about ~180 Mbp. The released sequence consists of the euchromatic portion, about 120 Mbp. The genome is distributed over five chromosomes: three large autosomes, a Y chromosome, and a fifth tiny chromosome containing only ~1 Mbp of euchromatin. The fly's 13 601 genes are approximately double the number in yeast, but are fewer than in *C. elegans*, perhaps a surprise. The average density of genes in the euchromatin sequence is 1 gene/9 kbp; much lower than the typical 1gene/kbp densities of prokaryotic genomes.

Despite the fact that insects are not very closely related to mammals, the fly genome is useful in the study of human disease. It contains homologues of 289 human genes implicated in various diseases, including cancer, and cardiovascular, neurological, endocrinological, renal, metabolic, and haematological diseases. Some of these homologues have different functions in humans and flies. Other human disease-associated genes can be introduced into, and studied in, the fly. For instance, the gene for human spinocerebellar ataxia type 3, when expressed in the fly, produces similar neuronal cell degeneration. There are now fly models for Parkinson disease and malaria.

The noncoding regions of the *D. melanogaster* genome must contain regions controlling spatio-temporal patterns of development. The developmental biology of the fly has been studied very intensively. It is therefore an organism in which the study of the genomics of development will prove extremely informative.

The Genome of Arabidopsis Thaliana

As a flowering plant *Arabidopsis thaliana* is a very distant relative of the other higher eukaryotic organisms for which genome sequences are available. It invites comparative analysis to identify common and specialised features. The *Arabidopsis thaliana* genome contains ~125 Mbp of DNA, of which 115.4 Mbp was reported in 2000 by The *Arabidopsis* Genome Initiative. There are five pairs of chromosomes, containing 25 498 predicted genes. The genome is relatively compact, with 1 gene/4.6 kbp on average. The genes of *Arabidopsis* are relatively small. Exons are typically 250 bp long, and introns relatively small, with mean length 170 bp. Typical of plant genes is an enrichment of coding regions in GC content.

Most *Arabidopsis* proteins have homologues in animals, but some systems are unique, among higher organisms, to plants, including cell wall production and photosynthesis. Many proteins shared with animals have diverged widely since the last common ancestor. Typical of another difference between plants and animals, 25% of the nuclear genes have signal sequences governing their transport into organelles—mitochondria and chloroplasts—compared to 5% of mitochon-drialtargeted nuclear genes in animals.

Traces of large- and small-scale duplication events appear in the *Arabidopsis* genome. 58% of the genome contains 24 duplicated segments, _ 100 kbp in length. The genes in the duplicated segments are usually *not* identifiable homologues. However, 4140 genes (17% of all genes) appear in 1528 tandem arrays of gene families, containing up to 23 members. The pattern of duplications suggests that an ancestor of *Arabidopsis* underwent a whole-genome duplication, to form a tetraploid species.

The Genome of Homo Sapiens (The Human Genome)

In February 2001, the International Human Genome Sequencing Consortium and Celera Genomics published, separately, drafts of the human genome. The sequence amounted to ~3.2 _ 109 bp, thirty times larger than the genomes of *C. elegans* or *D. melanogaster*. One reason for this disparity in size is that coding sequences form less than 5% of the human genome; repeat sequences over 50%. Perhaps the most surprising feature was the small number of genes identified. The finding of only about 20 000–25 000 genes suggests that alternative splicing patterns make a very significant contribution to our protein repertoire. It is estimated that ~50% of genes have alternative splicing patterns.

The published human genome is distributed over 22 chromosome pairs plus the X and Y chromosomes. The DNA contents of the autosomes range from 279 Mbp down to 48 Mbp. The X chromosome contains 163 Mbp and the Y chromosome only 51 Mbp. The exons of human protein-coding genes are relatively small compared to those in other known eukaryotic genomes. The introns are relatively long. As a result many protein-coding genes span long stretches of DNA. For instance, the dystrophin gene, coding for a 3685 amino acid protein, is _2.4 Mbp long.

Protein Coding Genes

Analysis of the human protein repertoire implied by the genome sequence has proved difficult because of the problems in reliably detecting genes, and alternative splicing patterns.

Repeat Sequences

Repeat sequences comprise over 50% of the genome:

- Transposable elements, or interspersed repeats—almost half the entire genome! These include the LINEs and SINEs.
- Retroposed pseudogenes
- Simple 'stutters'—repeats of short oligomers. These include the minisatellites anmicrosatellites. Trinucleotide repeats such as CAG, corresponding to glutamine repeats in the corresponding protein, are implicated in numerous diseases.
- Segmental duplications, of blocks of ~10–300 kbp. Interchromosomal duplications appear on nonhomologous chromosomes, sometimes at multiple sites. Some intrachromosomal duplications include closely spaced duplicated regions many kbp long of very similar sequence implicated in genetic diseases; for example Charcot-Marie-

Tooth syndrome type 1A, a progressive peripheral neuropathy resulting from duplication of a region containing the gene for peripheral myelin protein 22.

- Blocks of tandem repeats, including gene families

RNA

RNA genes in the human genome include:

1. 497 transfer RNA genes. One large cluster contains 140 tRNA genes within a 4Mbp region on chromosome 6.
2. Genes for 28S and 5.8S ribosomal RNAs appear in a 44 kbp tandem repeat unit of 150–200 copies. 5S RNA genes also appear in tandem arrays containing 200–300 genes, the largest of which is on chromosome 1.
3. Small nucleolar RNAs include two families of molecules that cleave and process ribosomal RNAs.
4. Splicesomal snRNAs, including the U1, U2, U4, U5, and U6 snRNAs, many owhich appear in clusters of tandem repeats of nearly-identical sequences, or inverted repeats.

Single-nucleotide Polymorphisms (SNPs)

A single-nucleotide polymorphism or SNP (pronounced 'snip') is a genetic variation between individuals, limited to a single base pair which can be substituted, inserted or deleted. Sickle-cell anaemia is an example of a disease caused by a specific SNP: an A → T mutation in the _-globin gene changes a Glu→ Val, creating a sticky surface on the haemoglobin molecule that leads to polymerization of the deoxy form.

SNPs are distributed throughout the genome, occurring on average every 5000 base pairs. Although they arose by mutation, many positions containing SNPs have low mutation rates, and provide stable markers for mapping genes. A consortium of four major academic genome centres and eleven private companies is creating a high-quality, high-density human SNP map, with a commitment to open public access. A recently published version contains 1.8 million SNPs. Not all SNPs are linked to diseases. Many are not within functional regions (although the density of SNPs is higher than the average in regions containing genes). Some SNPs that occur within exons are mutations to synonymous codons, or cause substitutions that do not significantly affect protein function. Other types of SNPs can cause more than local perturbation to a protein: (1) A mutation from a sense codon to a stop codon, or vice versa, will cause either premature truncation of protein synthesis or 'readthrough'. (2) A deletion or insertion will cause a phase shift in translation.

The A, B and O allelles of the genes for blood groups illustrate these possibilities. A and B alleles differ by four SNP substitutions. They code for related enzyme that add different saccharide units to an antigen on the surface of red blood cells.

The O allelle has undergone a mutation causing a phase shift, and produces no active enzyme. The red blood cells of type O individuals contain neither the A nor the B antigen. This is why people with type O blood are universal donors in blood transfusions. The loss of activity of the protein does not seem to carry any adverse consequences. Indeed, individuals of blood types B and O have greater resistance to smallpox.

Stong correlation of a disease with a specific SNP is advantageous in clinical work, because it is relatively easy to test for affected people or carriers. But if a disease arises from dysfunction of a specific protein, there ought to be many sites of mutations that could cause inactivation. A particular site may predominate if (1) all bearers of the gene are descendants of a single individual in whom the mutation occurred, and/or (2) the disease results from a *gain* rather than loss of a specific property, such as in the ability of sickle cell haemoglobin to polymerize, and/or (3) the mutation rate at a particular site is unusually high, as in the Gly380→Arg mutation in the fibroblast growth receptor gene FGFR3, associated with achondroplasia (a syndrome including short stature).

In contrast, many independent mutations have been detected in the BRCA1 and BRCA2 genes, loci associated with increased disposition to early-onset breast and ovarian cancer. The normal gene products function as tumour suppressors. Insertion or deletion mutants causing phase shifts generally produce a missing or inactive protein. But it cannot be deduced a priori whether a novel substitution mutant in BRCA1 or BRCA2 confers increased risk or not.

Treatments of diseases caused by defective or absent proteins include:

1. Providing normal protein We have mentioned insulin for diabetes, and Factor VIII in the most common type of

haemophilia. Another example is the administration of human growth hormone in patients with an absence or severe reduction in normal levels. Use of recombinant proteins eliminates the risk of transmission of AIDS through blood transfusions or of Creutzfeld-Jakob disease from growth hormone isolated from crude pituitary extracts.

2. Lifestyle adjustments that make the function unnecessary Phenylketoneuria (PKU) is a genetic disease caused by deficiency in phenylalanine hydroxylase, the enzyme that converts phenylalanine to tyrosine. Accumulation of high levels of phenylalanine causes developmental defects, including mental retardation. The symptoms can be avoided by a phenylalanine-free diet. Screening of newborns for high blood phenylalanine levels is legally required in the United States and many other countries.
3. Gene therapy to replace absent proteins is an active field of research. Other clinical applications of SNPs reflect correlations between genotype and reaction to therapy (pharmacogenomics). For example, a SNP in the gene for N-acetyl transferase (NAT–2) is correlated with peripheral neuropathy—weakness, numbness and pain in the arms, legs, hands or feet—as a side effect of treatment with isoniazid (isonicotinic acid hydrazide), a common treatment for tuberculosis. Patients who test positive for this SNP are given alternative treatment.

Genetic Diversity in Anthropology

SNP data are of great utility in anthropology, giving clues to historical variations in population size, and migration patterns. Degrees of genetic diversity are interpretable in terms of the size of the founding population. Founders are the original set of individuals from whom an entire population is descended.

Founders can be either original colonists, such as the Polynesians who first settled New Zealand, or merely the survivors of a nearextinction. Cheetahs show the effects of a population bottleneck, estimated to have occurred 10 000 years ago. All living cheetahs are as closely related to onanother as typical siblings in other species. Extrapolations of mitochondrial DNA variation in contemporary humans suggest a single maternal ancestor who lived 14 000–200 000 years ago. Calling her Eve suggests that she was the first woman.

But Mitochondrial Eve was in fact the founder of a surviving population following extinction of descendants of all her contemporaries.

Population-specific SNPs are informative about migrations. Mitochondrial sequences provide information about female ancestors and Y chromosome sequences provide information about male ancestors. For example, it has been suggested that the population of Iceland—first inhabited about 1100 years ago— is descended from Scandinavian males, and from females from both Scandinavia and the British Isles. Mediaeval Icelandic writings refer to raids on settlements in the British Isles.

A fascinating relationship between human DNA sequences and language families has been investigated by L. L. Cavalli-Sforza and colleagues. These studies have proved useful in working out interrelationships among American Indian languages. They confirm that the Basques, known to be a linguistically isolated population, have been genetically isolated also.

In the discovery of isolated populations, anthropological genetics provides data useful in medicine, for mapping disease genes is easier if the background variation is low. Genetically isolated populations in Europe include, in addition to the Basques, the Finns, Icelanders, Welsh, and Lapps.

Genetic Diversity and Personal Identification

Variations in our DNA sequences give us individual fingerprints, useful for identification and for establishment of relationships, including but not limited to questions of paternity. The use of DNA analysis as evidence in criminal trials is now well-established.

Genetic fingerprinting techniques were originally based on patterns of variable number tandem repeats (VNTRs), but have been extended to include analysis of other features including mitochondrial DNA sequences.

For most of us, all our mitochondria are genetically identical, a condition called homoplasmy. However, some individuals contain mitochondria with different DNA sequences; this is heteroplasmy. Such sequence variation in a disease gene can complicate the observed inheritance pattern of the disease.

The most famous case of heteroplasmy involved Tsar Nicholas II of Russia. After the revolution in 1917 the Tsar and his family were taken to exile in Yekaterinburg in Central Russia. During the night of 16–17 July 1918, the Tsar, Tsarina Alexandra, at least three of their five children, plus their physician and three servants who had accompanied the family, were killed, and their bodies buried in a secret grave. When the remains were rediscovered, assembly of the bones and examination of the dental work suggested—and sequence analysis confirmed—that the remains included an expected family group. The identity of the remains of the Tsarina were proved by matching the mitochondrial DNA sequence with that of a maternal relative, Prince Philip, Chancellor of the University of Cambridge, Duke of Edinburgh, and grandnephew of the Tsarina.

However, comparisons of mitochondrial DNA sequences of the putativremains of Nicholas II with those of two maternal relatives revealed a difference at base 16169: the Tsar had a C and the relatives a T. The extreme political and even religious sensitivities mandated that no doubts were tolerable. To confirm the identity beyond any reasonable question, the body of Grand Duke Georgij, brother of the Tsar, was exhumed.

Genetic Analysis of Cattle Domestication

Animal resources are an integral and essential aspect of human culture. Analysis of DNA sequences sheds light on their historical development and on the genetic variety characterizing modern breeding populations.

Contemporary domestic cattle include those familiar in Western Europe and North America, *Bos taurus*; and the zebu of Africa and India, *Bos indicus*. The most obvious difference in external appearance is the humpback of the zebu. It has been widely believed that the domestication of cattle occurred once, about 8000–10 000 years ago, and that the two species subsequently diverged.

Analysis of mitochondrial DNA sequences from European, African, and Asian cattle suggest, however, that (1) all European and African breeds are more closely related to each other than either is to Indian breeds, and (2) the two groups diverged about 200 000 years ago, implying recent independent domestications of different species. The similarity in physical appearance of African and Indian zebu (and other similarities at the molecular level; for instance, VNTR markers in nuclear DNA) must then be attributable to importation of male cattle from India to East Africa.

Chapter 8

Genomics and Global Governance

Bioinformatics

Bioinformatics is the application of statistics and computer science to the field of molecular biology.

The term *bioinformatics* was coined by Paulien Hogeweg and Ben Hesper in 1978 for the study of informatic processes in biotic systems. Its primary use since at least the late 1980s has been in genomics and genetics, particularly in those areas of genomics involving large-scale DNA sequencing.

Bioinformatics now entails the creation and advancement of databases, algorithms, computational and statistical techniques and theory to solve formal and practical problems arising from the management and analysis of biological data.

Over the past few decades rapid developments in genomic and other molecular research technologies and developments in information technologies have combined to produce a tremendous amount of information related to molecular biology. It is the name given to these mathematical and computing approaches used to glean understanding of biological processes.

Common activities in bioinformatics include mapping and analysing DNA and protein sequences, aligning different DNA and protein sequences to compare them and creating and viewing 3-D models of protein structures. The primary goal of bioinformatics is to increase the understanding of biological processes. What sets it apart from other approaches, however, is its focus on developing and applying computationally intensive techniques (e.g., pattern recognition, data mining, machine learning algorithms, and visualization) to achieve this goal.

Major research efforts in the field include sequence alignment, gene finding, genome assembly, drug design, drug discovery, protein structure alignment, protein structure prediction, prediction of gene expression and protein-protein interactions, genome-wide association studies and the modeling of evolution.

Bioinformatics was applied in the creation and maintenance of a database to store biological information at the beginning of the "genomic revolution", such as nucleotide and amino acid sequences. Development of this type of database involved not only design issues but the development of complex interfaces whereby researchers could both access existing data as well as submit new or revised data.

In order to study how normal cellular activities are altered in different disease states, the biological data must be combined to form a comprehensive picture of these activities.

Therefore, the field of bioinformatics has evolved such that the most pressing task now involves the analysis and interpretation of various types of data, including nucleotide and amino acid sequences, protein domains, and protein structures. The actual process of analysing and interpreting data is referred to as computational biology. Important sub-disciplines within bioinformatics and computational biology include:

- the development and implementation of tools that enable efficient access to, and use and management of, various types of information.
- the development of new algorithms (mathematical formulas) and statistics with which to assess relationships among members of large data sets, such as methods to locate a gene within a sequence, predict protein structure and/or function, and cluster protein sequences into families of related sequences.

There are two fundamental ways of modelling a Biological system (e.g. living cell) both coming under Bioinformatic approaches.

- Static
    - o Sequences - Proteins, Nucleic acids and Peptides
    - o Structures - Proteins, Nucleic acids, Ligands (including metabolites and drugs) and Peptides
    - o Interaction data among the above entities including microarray data and Networks of proteins, metabolites.

- Dynamic
    - o Systems Biology comes under this category including reaction fluxes and variable concentrations of metabolites
    - o Multi-Agent Based modelling approaches capturing cellular events such as signalling, transcription and reaction dynamics.

A broad sub-category under bioinformatics is structural bioinformatics.

Major Research Areas

Sequence Analysis

Since the Phage Φ-X174 was sequenced in 1977, the DNA sequences of thousands of organisms have been decoded and stored in databases. This sequence information is analysed to determine genes that encode polypeptides (proteins), RNA genes, regulatory sequences, structural motifs, and repetitive sequences. A comparison of genes within a species or between different species can show similarities between protein functions, or relations between species (the use of molecular systematics to construct phylogenetic trees). With the growing amount of data, it long ago became impractical to analyze DNA sequences manually. Today, computer programs such as BLAST are used daily to search sequences from more than 260 000 of organisms, containing over 190 billion nucleotides. These programs can compensate for mutations (exchanged, deleted or inserted bases) in the DNA sequence, in order to identify sequences that are related, but not identical. A variant of this sequence alignment is used in the sequencing process itself.

The so-called shotgun sequencing technique (which was used, for example, by The Institute for Genomic Research to sequence the first bacterial genome, *Haemophilus influenzae*) does not produce entire chromosomes, but instead generates the sequences of many thousands of small DNA fragments (ranging from 35 to 900 nucleotides long, depending on the sequencing technology). The ends of these fragments overlap and, when aligned properly by a genome assembly program, can be used to reconstruct the complete genome. Shotgun sequencing yields sequence data quickly, but the task of assembling the fragments can be quite complicated for larger genomes. For a genome as large as the human genome, it may take many days of CPU time on large-memory, multiprocessor computers to assemble the fragments, and the resulting assembly will usually contain numerous gaps that have to be filled in later. Shotgun sequencing is the method of choice for virtually all genomes sequenced today, and genome assembly algorithms are a critical area of bioinformatics research.

Another aspect of bioinformatics in sequence analysis is annotation, which involves computational gene finding to search for protein-coding genes, RNA genes, and other functional sequences within a genome. Not all of the nucleotides within a genome are part of genes. Within the genome of higher organisms, large parts of the DNA do not serve any obvious purpose. This so-called junk DNA may, however, contain unrecognized functional elements. Bioinformatics helps to bridge the gap between genome and proteome projects — for example, in the use of DNA sequences for protein identification.

Genome Annotation

In the context of genomics, annotation is the process of marking the genes and other biological features in a DNA sequence. The first genome annotation software system was designed in 1995 by Dr. Owen White, who was part of the team at The Institute for Genomic Research that sequenced and analysed the first genome of a free-living organism to be decoded, the bacterium *Haemophilus influenzae*. Dr. White built a software system to find the genes (places in the DNA sequence that encode a protein), the transfer RNA, and other features, and to make initial assignments of function to those genes. Most current genome annotation systems work similarly, but the programs available for analysis of genomic DNA are constantly changing and improving.

Computational Evolutionary Biology

Evolutionary biology is the study of the origin and descent of species, as well as their change over time. Informatics has assisted evolutionary biologists in several key ways; it has enabled researchers to:

- trace the evolution of a large number of organisms by measuring changes in their DNA, rather than through physical taxonomy or physiological observations alone,
- more recently, compare entire genomes, which permits the study of more complex evolutionary events, such as gene

duplication, horizontal gene transfer, and the prediction of factors important in bacterial speciation,

- build complex computational models of populations to predict the outcome of the system over time
- track and share information on an increasingly large number of species and organisms.

Future work endeavours to reconstruct the now more complex tree of life.

The area of research within computer science that uses genetic algorithms is sometimes confused with computational evolutionary biology, but the two areas are not necessarily related.

Analysis of Gene Expression

The expression of many genes can be determined by measuring mRNA levels with multiple techniques including microarrays, expressed cDNA sequence tag (EST) sequencing, serial analysis of gene expression (SAGE) tag sequencing, massively parallel signature sequencing (MPSS), or various applications of multiplexed in-situ hybridization.

All of these techniques are extremely noise-prone and/or subject to bias in the biological measurement, and a major research area in computational biology involves developing statistical tools to separate signal from noise in high-throughput gene expression studies. Such studies are often used to determine the genes implicated in a disorder: one might compare microarray data from cancerous epithelial cells to data from non-cancerous cells to determine the transcripts that are up-regulated and down-regulated in a particular population of cancer cells.

Analysis of Regulation

Regulation is the complex orchestration of events starting with an extracellular signal such as a hormone and leading to an increase or decrease in the activity of one or more proteins. Bioinformatics techniques have been applied to explore various steps in this process.

For example, promoter analysis involves the identification and study of sequence motifs in the DNA surrounding the coding region of a gene. These motifs influence the extent to which that region is transcribed into mRNA. Expression data can be used to infer gene regulation: one might compare microarray data from a wide variety of states of an organism to form hypotheses about the genes involved in each state. In a single-cell organism, one might compare stages of the cell cycle, along with various stress conditions (heat shock, starvation, etc.). One can then apply clustering algorithms to that expression data to determine which genes are co-expressed. For example, the upstream regions (promoters) of co-expressed genes can be searched for over-represented regulatory elements.

Analysis of Protein Expression

Protein microarrays and high throughput (HT) mass spectrometry (MS) can provide a snapshot of the proteins present in a biological sample. Bioinformatics is very much involved in making sense of protein microarray and HT MS data; the former approach faces similar problems as with microarrays targeted at mRNA, the latter involves the problem of matching large amounts of mass data against predicted masses from protein sequence databases, and the complicated statistical analysis of samples where multiple, but incomplete peptides from each protein are detected.

Analysis of Mutations in Cancer

In cancer, the genomes of affected cells are rearranged in complex or even unpredictable ways. Massive sequencing efforts are used to identify previously unknown point mutations in a variety of genes in cancer. Bioinformaticians continue to produce specialized automated systems to manage the sheer volume of sequence data produced, and they create new algorithms and software to compare the sequencing results to the growing collection of human genome sequences and germline polymorphisms.

New physical detection technologies are employed, such as oligonucleotide microarrays to identify chromosomal gains and losses (called comparative genomic hybridization), and single-nucleotide polymorphism arrays to detect known *point mutations*. These detection methods simultaneously measure several hundred thousand sites throughout the genome, and when used in high-throughput to measure thousands of samples, generate terabytes of data per experiment. Again the massive amounts and new types of data generate new opportunities for bioinformaticians. The data is often found to contain considerable variability, or noise, and thus Hidden Markov model and change-point analysis methods are being developed to infer real copy number changes.

Another type of data that requires novel informatics development is the analysis of lesions found to be recurrent among many tumours.

Comparative Genomics

The core of comparative genome analysis is the establishment of the correspondence between genes (orthology analysis) or other genomic features in different organisms. It is these intergenomic maps that make it possible to trace the evolutionary processes responsible for the divergence of two genomes. A multitude of evolutionary events acting at various organizational levels shape genome evolution. At the lowest level, point mutations affect individual nucleotides. At a higher level, large chromosomal segments undergo duplication, lateral transfer, inversion, transposition, deletion and insertion. Ultimately, whole genomes are involved in processes of hybridization, polyploidization and endosymbiosis, often leading to rapid speciation. The complexity of genome evolution poses many exciting challenges to developers of mathematical models and algorithms, who have recourse to a spectra of algorithmic, statistical and mathematical techniques, ranging from exact, heuristics, fixed parameter and approximation algorithms for problems based on parsimony models to Markov Chain Monte Carlo algorithms for Bayesian analysis of problems based on probabilistic models.

Many of these studies are based on the homology detection and protein families computation.

Modeling Biological Systems

Systems biology involves the use of computer simulations of cellular subsystems (such as the networks of metabolites and enzymes which comprise metabolism, signal transduction pathways and gene regulatory networks) to both analyze and visualize the complex connections of these cellular processes. Artificial life or virtual evolution attempts to understand evolutionary processes via the computer simulation of simple (artificial) life forms.

High-throughput Image Analysis

Computational technologies are used to accelerate or fully automate the processing, quantification and analysis of large amounts of high-information-content biomedical imagery. Modern image analysis systems augment an observer's ability to make measurements from a large or complex set of images, by improving accuracy, objectivity, or speed.

A fully developed analysis system may completely replace the observer. Although these systems are not unique to biomedical imagery, biomedical imaging is becoming more important for both diagnostics and research. Some examples are:

- high-throughput and high-fidelity quantification and sub-cellular localization (high-content screening, cytohistopathology, Bioimage informatics)
- morphometrics
- clinical image analysis and visualization
- determining the real-time air-flow patterns in breathing lungs of living animals
- quantifying occlusion size in real-time imagery from the development of and recovery during arterial injury
- making behavioural observations from extended video recordings of laboratory animals
- infrared measurements for metabolic activity determination
- inferring clone overlaps in DNA mapping, e.g. the Sulston score.

Structural Bioinformatic Approaches

Prediction of Protein Structure

Protein structure prediction is another important application of bioinformatics. The amino acid sequence of a protein, the so-called primary structure, can be easily determined from the sequence on the gene that codes for it. In the vast majority of cases, this primary structure uniquely determines a structure in its native environment. (Of course, there are exceptions, such as the bovine spongiform encephalopathy - aka Mad Cow Disease - prion.) Knowledge of this structure is vital in understanding the function of the protein. For lack of better terms, structural information is usually classified as one of *secondary*, *tertiary* and *quaternary* structure. A viable general solution to such predictions remains an open problem. As of now, most efforts have been directed towards heuristics that work most of the time.

One of the key ideas in bioinformatics is the notion of homology. In the genomic branch of bioinformatics, homology is used to predict the function of a gene: if the sequence of gene *A*, whose function is known, is homologous to the sequence of gene *B,* whose function is unknown, one could infer that B may share A's function. In the structural branch of bioinformatics, homology is used to determine which parts of a protein are important in structure formation and interaction with other proteins. In a technique called homology modeling, this information is used to predict the structure of a protein once the structure of a homologous protein is known. This currently remains the only way to predict protein structures reliably. One example of this is the similar protein homology between hemoglobin in humans and the hemoglobin in legumes (leghemoglobin).

Both serve the same purpose of transporting oxygen in the organism. Though both of these proteins have completely different amino acid sequences, their protein structures are virtually identical, which reflects their near identical purposes. Other techniques for predicting protein structure include protein threading and *de novo* (from scratch) physics-based modeling.

Molecular Interaction

Efficient software is available today for studying interactions among proteins, ligands and peptides. Types of interactions most often encountered in the field include -Protein-ligand (including drug), protein-protein and protein-peptide. Molecular dynamic simulation of movement of atoms about rotatable bonds is the fundamental principle behind computational algorithms, termed docking algorithms for studying molecular interactions.

Docking Algorithms

In the last two decades, tens of thousands of protein three-dimensional structures have been determined by X-ray crystallography and Protein nuclear magnetic resonance spectroscopy (protein NMR). One central question for the biological scientist is whether it is practical to predict possible protein-protein interactions only based on these 3D shapes, without doing protein-protein interaction experiments. A variety of methods have been developed to tackle the Protein-protein docking problem, though it seems that there is still much work to be done in this field.

Software and Tools

Software tools for bioinformatics range from simple command-line tools, to more complex graphical programs and standalone web-services available from various bioinformatics companies or public institutions.

Web Services in Bioinformatics

SOAP and REST-based interfaces have been developed for a wide variety of bioinformatics applications allowing an application running on one computer in one part of the world to use algorithms, data and computing resources on servers in other parts of the world. The main advantages derive from the fact that end users do not have to deal with software and database maintenance overheads.

Basic bioinformatics services are classified by the EBI into three categories: SSS (Sequence Search Services), MSA (Multiple Sequence Alignment) and BSA (Biological Sequence Analysis). The availability of these service-oriented bioinformatics resources demonstrate the applicability of web based bioinformatics solutions, and range from a collection of standalone tools with a common data format under a single, standalone or web-based interface, to integrative, distributed and extensible bioinformatics workflow management systems.

Bibliography

Alberts B, Johnson A: *Molecular Biology of the Cell*, New York: Garland Science, 2002.

Altered Fates: *Gene Therapy and the Retooling of Human Life*, W. W. Norton and Company, New York, 1995.

Ayala, F. J.: *The Genetic Structure of Populations*, W.H. Freeman & Co., San Francisco, California, 1977.

Bajema, C. J.: *Natural Selection in Human Populations, the Measurement of Ongoing Genetic Evolution in Contemporary Societies*: New York, Wiley, 1971.

Balkwill, Dr. Fran: *DNA Is Here to Stay*, Carolrhoda Books, Inc., Minneapolis, 1993.

Balkwill, Fran: *DNA Is Here to Stay*, Carolrhoda Books, Inc., Minneapolis, 1993.

Banerjee, S : *Cell Biochemistry*, Dominant Pub, Delhi, 2009.

Bast, R.C. Jr.: *Cancer Medicine*, Decker, Hamilton, 2000.

Beckner, M.: *The Biological Way of Thought*, University of California Press, Berkeley and Los Angeles, 1968.

Berg, Paul, and Singer, Maxine: *Dealing with Genes: The Language of Heredity. Mill Valley*, University Science Books, CA, 1992.

Bernard, C.: *An Introduction to the Study of Experimental Medicine*, Dover Publications, New York, 1957.

Birch, L. C.: *The Distribution and Abundance of Animals*, University of Chicago Press, Chicago, Illinois, 1954.

Bradley, W.G.: *Neurology in Clinical Practice*, Butterworth Heinemann, Boston, 2000.

Brookes, Martin: *Get a Grip on Genetics, Time Life Books*, East Sussex, England, 1998.

Cavalli-Sforza, L.L., P. Menozzi: *The History and Geography of Human Genes*. Princeton: Princeton University Press, 1994.

Credi, A.: *Molecular Devices and Machines: a Journey into the Nano World*, Wiley-VCH, Weinheim, 2003.

Curtis, M.: *The Geometry of DNA: A Structural Revision*. London: Blue Gallery, 1996.

Dabholkar, A.R. : *General Plant Breeding*, Concept, Delhi, 2006.

Dando, M.: *Biological Warfare in the 21st Century: Biotechnology and the Proliferation of Biological Weapons*. London: Brassey's, 1994.

Darlington, C. D., and Mather, K.: *The Elements of Genetics*, London, Allen and Unwin, 1949.

David M.: *Basic Principles and Calculations in Chemical Engineering*, Prentice-Hall, 1996.

Dawkins, R.: *The Selfish Gene*. Oxford: Oxford University Press, 1976.

DeVita, V.T.: *Cancer: Principles and Practice of Oncology*, Lippincott, Philadelphia, 2005.

Diaz, E.: *Microbial Biodegradation: Genomics and Molecular Biology*, Caister Academic Press, UK, 2008.

Dobzhanshy, T.: *Genetics of the Evolutionary Process*, Columbia University Press, New York, 1970.

Easty, D. M.: *Cell Biology: Reading*, Addison- Wesley, Mass, 1970.

Fisher, R. A.: *The Genetical Theory of Natural Selection*, Clarendon Press, Oxford, 1930.

Fritz, A.: *International Classification of Diseases for Oncology*, World Health Organization, Geneva, 2000.

Fukuyama F.: *Our Posthuman Future: Consequences of the Biotechnology Revolution*, Farrar, Straus and Giroux, NewYork, 2002.

Gall, Joseph G.: *Landmark Papers in Cell Biology*, Cold Spring Harbor Laboratory Press, Plainview, NY, 2001.

George S. Paul: *Beyond Humanity: Cyber Evolution and Future Minds*, Charles River Media, Roackland, 1996.

Gillis, Justin: *Drug Firms, Gene Labs to Map Genetic Code*, The Daily News, Longview, WA, 1999.

Glover, D. M., and Hames, B. D.: *Genes and Embryos*: New York, Oxford University Press, 1989.

Glut, D.F.: *The Frankenstein Legend: A Tribute to Mary Shelley and Boris Karloff*. Metachen, New Jersey: The Scarecrow Press, 1973.

Griffiths A.J.F.: *Introduction to Genetic Analysis*, W.H. Freeman and Company, New York, USA, 2005.

Groom, M.J., Meffe, G.K. : *Principles of Conservation Biology*, Sunderland, MA: Sinauer Associates, 2006.

Guyton, A.C.; Hall, J.E.: *Textbook of Medical Physiology*, Saunders, Philadelphia, 2006.

Halacy, D.S., Jr.: *Genetic Revolution, Shaping Life for Tomorrow*, Harper & Row, Publishers, New York, 1974.

Hamerton, J. L.: *Human Cytogenetics*, Academic Press, New York, 1971.

Hanley, R.: *Is Data Human? The Metaphysics of Star Trek*, Boxtree, London, 1998.

Hartl D, Jones E: *Genetics: Analysis of Genes and Genomes,* Jones & Bartlett, 2005.

Hartl, Daniel L.: *Basic Genetics*, Jones and Bartlett Publishers, Boston, 1991.

Haubrich, W.S.: *Bockus Gastroenterology*, Saunders, Philadelphia, 1995.

Heider, J. and Rabus, R: *Microbial Biodegradation: Genomics and Molecular Biology*, Caister Academic Press, UK, 2008.

Herrero A and Flores E,: *The Cyanobacteria: Molecular Biology, Genomics and Evolution* (1st ed.). Caister Academic Press, 2008.

Holtz, Robert D.; William, Kovacs D.: *An Introduction to Geotechnical Engineering*, Prentice Hall, UK, 1981.

Jangir, M.L. : *Cell Biology : Fundamentals and Applications*, Agrobios, Delhi, 2009.

Jonoska, N.: *Self-Assembling DNA Graphs, DNA-Based Computers VIII*, Springer-Verlag, Berlin, 2003.

Kelves, D.: *In the Name of Eugenics: Genetics and the Uses of Human Heredity*. Harmondsworth: Penguin.

King, R.C.; Stansfield, W.D.: *A Dictionary of Genetics*, Oxford Univ. Press, New York, 2002.

Klotz, J. W.: *Genes, Genesis, and Evolution*: St. Louis, Mo., Concordia Publishing Co., 1970.

Landa, Norbert: *Ingenious Genes Microexplorers*, Barron's, Hauppauge, New York, 1997.

Lewis, Ricki: *Human Genetics: Concepts and Applications,* Boston, Mass: McGraw Hill. 1994.

Lewontin, R.C.: *The Doctrine of DNA: Biology as Ideology*. Harmondsworth: Penguin, 1993.

Lindee, M.S. : *The DNA Mystique: The Gene as a Cultural Icon*. New York: W.H. Freeman, 1995.

Luca Cavalli-Sforza, Luigi: *The History and Geography of Human Genes*. Princeton University Press, 1996.

Lumsden C., and E. Wilson: *Genes, Mind and Culture: The Coevolutionary Process*. Cambridge, MA: Harvard University Press, 1981.

Lyon, Jeff and Gorner, Peter. Altered Fates: *Gene Therapy and the Retooling of Human Life*, W. W. Norton and Company, New York, 1995.

Margulis, L.: *Symbiosis in Cell Evolution*. San Francisco: W.H. Freeman, 1981.

Mayr, E.: *Change of Genetic Environment and Evolution*, Allen and Unwin, London, 1954.

Michael R.: Cummings. *Essentials of Genetics*, Prentice Hall, New Jersey, 1996.

Migloni, G.S.: *Dictionary of Plant Genetics and Molecular Biology*, Hawthorne Press, New York, 1998.

Mishra, Shubhrata R. : *Textbook of Mycology*, Discovery Publishing House, Delhi, 2010.

Nottingham, S.F.: *Eat Your Genes: How Genetically Modified Food is Entering Our Diet*, Zed Books, London, 1998.

Prakash, Chandra Prakash : *Cell Physiology*, Discovery, Delhi, 2010.

Rietman, Ed.: *Molecular Engineering of Nanosystems*, Springer, New York, 2001.

Robin J. Heyden: *Biology: Exploring Life*, Pearson Prentice Hall, Boston, Massachusetts, 2006.

Robin Whyman: *Applied Organometallic Chemistry and Catalysis*, Oxford University Press, Delhi, 2005.

Roobeek, A: *The Biotechnology Revolution?*, Blackwell, Oxford, 1995.

Rorvik, D.M.: *In His Image: The Cloning of a Man*. New York: J.B. Lippincott, 1978.

Scriver, C.R.: *The Metabolic and Molecular Basis of Inherited Disease*, McGraw Hill, New York, 2001.

Stebbins, G. L.: *Darwin to DNA, Molecules to Humanity*, W. H. Freeman, San Francisco, 1982.

Steen, R.G.: *DNA and Destiny: Nature and Nurture in Human Behaviour*. New York and London: Plenum Press, 1996.

Sturtevant, Alfred: *History of Genetics*, Harper and Row, New York, 1965.

Stwertka, Eve and Albert: *Genetic Engineering*, Franklin Watts, New York, 1982.

Suzuki, D., and Knudtson, P.: *Genethics: The Clash Between the New Genetics and Human Values*, Harvard University Press, Cambridge, Mass, 1989.

Vasudev Bhatnagar: *Cell Science and Technology*, Campus Books, Delhi, 2009.

Watson, J. D.: *The Double Helix*, Antheneum, New York, 1968.

Wilson, E. O., and Lumden, C.: *Genes, Mind, and Culture: The Evolutionary Process*, Harvard University Press, Cambridge, Mass, 1981.

Wilson, E. O., and Lumden, C.: *Genes, Mind, and Culture: The Evolutionary Process*: Cambridge, Mass., Harvard University Press, 1981.

Wilson, I.: *The Blood and The Shroud*. London: Weidenfeld & Nicolson, 1998.

Winston, R.: *The Future of Genetic Manipulation*, Phoenix, London, 1997.

Wright, L.: *Twins: Genes, Environment and the Mystery of Human Identity*. London: Weidenfeld & Nicolson, 1997.

Index

I

L

M

N

O

P

R

S

T

U

V

OOO